the Water Problem

Climate Change and Water Policy in the United States

Pat Mulroy

Editor

BROOKINGS INSTITUTION PRESS
Washington, D.C.

Copyright © 2017
THE BROOKINGS INSTITUTION
1775 Massachusetts Avenue, N.W., Washington, D.C. 20036
www.brookings.edu

The Brookings Institution is a private nonprofit organization devoted to research, education, and publication on important issues of domestic and foreign policy. Its principal purpose is to bring the highest quality independent research and analysis to bear on current and emerging policy problems. Interpretations or conclusions in Brookings publications should be understood to be solely those of the authors.

Library of Congress Cataloging-in-Publication data are available.
ISBN 978-0-8157-2784-2 (pbk : alk. paper)
ISBN 978-0-8157-2786-6 (ebook)

9 8 7 6 5 4 3 2 1

Typeset in Jenson Pro

Composition by Westchester Publishing Services

Contents

Foreword

BRUCE BABBITT

As this book goes to press, the great California drought continues to dominate news from the West. The stories proliferate: crops withering, fish dying, wells going dry, lakes turning to dust, fires on the land, and water rationing—all prompting some commentators to suggest that California has seen its best days and now enters an era of economic decline.

Visions of an impending hydrologic apocalypse make for good copy. The reality, however, is not so much an absolute lack of water, as a deficit of planning and good management and political leadership.

California, like most of the West, has a long history of profligate water use, fragmented efforts to build and manage water infrastructure, and an absence of integrated planning. To make matters worse, we all too often assume that water is free, that we can consume it without limit, and if we need more, someone else will pay for it.

Our faltering efforts at water management are also blighted by a legalistic, adversarial political culture that tends to define water rights as absolute entitlements to be used or misused at will, ignoring impacts on neighbors and downstream users.

My state of Arizona exemplifies the tradition of adversarial water politics. I recall one controversy in which our governor sent the 158th Infantry Regiment of the Arizona National Guard to the Colorado River in a futile attempt to block construction of a dam that would divert water to Los Angeles.

Old-style water wars are by no means a thing of the past. The state of Colorado still divides east versus west over transmountain diversions. California continues to quarrel over a critical tunnel project to divert more water from the Sacramento River to southern California. Georgia, Alabama, and Florida are escalating their fight over the waters of the Apalachicola-Chattahoochee River Basin up to the Supreme Court.

Notwithstanding this history of contention, there are encouraging signs that a new era of cooperation is at last beginning to emerge. This transition toward a future of cooperation is what unites the essays and underlines the importance of this book. While treating distinct issues in varied regions, these essays all point in new directions, challenging us to consider working together as a better pathway into the future.

This emerging spirit is now evident even on the much disputed and litigated Colorado River. Nevada is pioneering water banking behind Hoover Dam and recharging water into ground water basins across the state line in Arizona. The Lower Basin states have joined together to deliver water to Mexico to restore the fabled wetlands of the Colorado River Delta. Nevada and California share water saved through investments in efficiency.

Spurred to action by the drought emergency, California is beginning to lead in innovation. Governor Jerry Brown has announced a California Water Action Plan that touches on many of the issues discussed in this book. An organizing theme is the urgent need for intergovernmental collaboration to replace conflicting and piecemeal efforts. And a related theme running through the California plan is the imperative to address all sides of the water equation by both increasing supply and reducing demand.

On the supply side, the most neglected issue is management of groundwater. This resource, out of sight and mostly out of mind, supplies nearly 40 percent of the water used in California (and 20 percent in the United States). The Ogallala aquifer, extending beneath eight states in the high plains, is surely an appropriate place to begin thinking anew about the management and replenishment of aquifers that should not be left to disappear in an unconstrained race to the bottom.

We must also utilize and integrate all potential water sources, including recycling, groundwater recharge, storm water capture, and desalination. Conservation and efficient use are still the low-hanging fruits to be harvested. California, for example, has in 2015 reduced urban water consumption by more than 20 percent in response to the drought. Our task now is to embed these conservation measures into regular daily life, not just in times of extreme drought.

The proper pricing of water, including tiered water rates, remains an underutilized tool for water management. Water markets can provide an effective method to price and allocate water.

The onset of climate change—bringing droughts, extreme storm events, and sea-level rise—will call for more and better data, more robust science, and integrated planning. The imperative to protect endangered species and manage entire watersheds and ecosystems for their environmental values will add to the challenges we face.

In the old days of adversary, zero-sum water management lawyers were the dominant players among the insider "water buffaloes" who fought and made water policy, mostly out of public view. In the new world envisioned in this work, the making of water policy will expand to include many more hydrologists, agronomists, biologists, engineers, and economists. And the circle must also widen to include a lot more public understanding and attention to the issues discussed in this excellent book.

Introduction
The Hidden Obstacles to Adaptation

PAT MULROY

I will never forget the warm spring morning in 2002 when residents of southern Nevada were shaken by the reality that their assumed reliable fifty-year water supply had evaporated. After two years of a "normal" and manageable drought event in the Colorado River system, the unimaginable was enfolding. Kay Brothers, Southern Nevada Water Authority's engineering and operations leader, walked into the regular Monday morning executive meeting looking more concerned than I had ever seen her. She had just received the news from the federal Bureau of Reclamation that the anticipated runoff from the Rocky Mountains into Lake Powell that year was going to be only 25 percent of normal—an event without precedent. The reservoir system on the river was facing dramatic declines.

Southern Nevada had spent the entire previous decade in difficult discussions with its neighbors on the river, which had resulted in the region being afforded a longer-term water supply. However, the arrangement relied on a relatively stable and healthy river system. The region was just beginning to mature economically, the way Phoenix had in the 1970s and Southern California had in the decades before that. Southern Nevada had relied on the bargains it had struck with its neighbors and was already using more river water annually than that to which it was entitled. If the reservoirs became seriously stressed, that would no longer be possible—southern Nevada would essentially run out of water. On that early spring day, water resource planning in southern Nevada changed forever.

By 2007 the Colorado River basin was not the only area beginning to feel the strain. That year representatives of seven large urban water agencies sequestered themselves at San Francisco's facilities at Hetch Hetchy Reservoir to explore with representatives of the science community the impacts of a changing climate on their part of the United States. Participants included Seattle, Portland, San Francisco, San Diego, the Metropolitan Water District of Southern California, Las Vegas, Denver, and New York. They realized and accepted that dramatically altered weather patterns could result in consequences for which they had no strategies or solutions, and the lack of political will required to implement what few solutions might appear was daunting. The rule of thumb for the utilities had always been that past experience would guide the parameters of the future. Utilities managers looked in the rearview mirror, studied historical weather events, and designed systems that would accommodate those probabilities. With no hard data on which to rely, utility and resource planners were left working with only the cone of uncertainty—a range of probabilities that would narrow only with time.

In the years since that 2007 gathering, much has changed. Temperatures have continued to rise and extreme weather events have become the norm. For the most part, American society has taken water for granted, assuming it would always be available in whatever volumes were required and with a quality that was completely safe to consume. The United States has been in the enviable global minority in this respect: water has poured out of American faucets reliably twenty-four hours a day, and disposal and treatment of wastewater have occurred in the normal course of urban existence. However, the need to adapt to dramatically altered circumstances could not be more acute. Changing climate patterns may be the most striking manifestation, but climate change exists in a confluence of global changes, all of which dangerously intersect and compound the water resource challenges.

In the last several assessment cycles, the World Economic Forum (WEF) has ranked inadequate water supplies as the global risk with the greatest consequences over the next decade. Global populations are expected to reach or exceed 9 billion by the year 2050. To feed this explosion of humanity, the WEF estimates that 50–60 percent more food needs to be produced by 2030 and that energy production will require 80 percent more water than it consumes today. As economically developing countries raise their national standard of living, their diets change, driving more U.S. food exports. Already Saudi Arabia owns farms and ranches in the southwestern United States, and billions of dollars' worth of agricultural products are exported to China and Southeast Asia every year. Increased agricultural and energy demands increase the demands for water globally.

At the same time, the pressures on U.S. urban hydrological systems are mounting. Urban economies are growing as people increasingly move to cities for jobs, and aging water and wastewater systems are suffering under the strain of decades of deferred maintenance. The leakage rates in some larger, older cities can go as high as 40 percent, and the tragedy of Flint, Michigan, makes it impossible to ignore the need for a laser focus on refurbishing U.S. infrastructure. When systems already weakened by age and inattention are subjected to extreme weather events, they cannot withstand the strain. Storm sewers overflow and treatment plants are overwhelmed, compromising water quality.

This book examines the impact of a changing climate on the resilience of U.S. water supplies and utilities by exploring several instances of critical water management policy drawn from across the country. Most of these case studies and vignettes show the degree to which utility and resource planning are fundamentally changing. Even as communities struggle to adapt, certain policy, regulatory, and governance issues are emerging that must be addressed.

On the resource side, utilities managers are shifting from a traditional reactive stance to one that is far more proactive. That is particularly visible in the drought-ravaged western United States. In the past, when a drought hit, the doctrine of prior appropriation took over, and water was simply withheld from the junior water rights holders, benefiting and protecting those who had acquired their rights earlier in time. In urban utilities, draconian emergency drought measures are adopted, such as limiting outside watering or, in extreme circumstances, limiting individual household use.

Never before has a region collectively taken measures before disaster struck to bolster reservoir systems and prevent the worst from happening. The length and severity of twenty-first-century droughts and the changes that have occurred in the West, both demographic and economic, have cast doubt on the value of the prior appropriative water rights doctrine. Rather than this century-old system being cast aside legislatively, however, negotiated arrangements between users can afford even the most senior water rights holders far greater surety.

Another notion that is shifting is that water is local. It is managed and administered locally, and communities develop an enormous sense of ownership around the resource. That may still be the case for those fortunate enough to have as a water source a discrete groundwater basin with no other users, but when a river, lake, or groundwater basin is a shared source, the local dimension fades rapidly. The diminishment of a local sense of ownership is even more pronounced in the West, where naturally discrete watersheds have been connected through aqueducts and diversions, creating large interdependent areas. Climate events affect the entire region, necessitating regional planning, regional management, and a reexamination of shared risks. Resilience is to be found in strategic proactive partnerships among those who bear the risk. Although such arrangements are usually fraught with political challenges, the advantages of taking the larger regional view far outweigh the trials of the narrower political challenges. Larger multistate conservation measures protect a shared system from collapse and lessen the economic and social impacts that will inevitably occur if the situation is left unaddressed.

The recent widespread effects of the protracted drought in California loom large as an example of how regional divisiveness can threaten to bring an entire state to its knees. The years leading up to the drought are a story of missed opportunities. Years of overly abundant precipitation were wasted; none of the water was allowed to move out of the region to existing reservoirs and groundwater basins to create a reserve for leaner resource years. The operation of the pumps on the State Water Project delivering water to Southern California and the Central Valley had been dramatically curtailed by the court to save fish species in the Delta. Parochial politics prevented the development of a longer-term view and a proactive approach to managing the shared water system of the state. And, sadly, the ostensible reason for this paralysis, saving the fish of the Sacramento-San Joaquin River Delta region, wasn't accomplished either.

Dramatically changing weather patterns are not only affecting water resource management in California and the Western United States. Scientists have not suggested that the planet as a whole will experience less precipitation, rather that where it falls and how it falls will change. Some areas will see snow replaced by rain. Some will trend drier and some will trend wetter. Today's infrastructure and resource planning strategies were developed under a very different set of assumptions. This becomes particularly visible in flood-ravaged zones. Year after year communities and cropland are destroyed by floods, which are only becoming more devastating. Here too, questions about the adequacy of water management systems to withstand such catastrophic events emerge, followed by the challenge of reimagining such systems. Can new flood-control systems be designed that move water out of the affected areas and into zones plagued by drought? Can such a design recharge groundwater basins? Can it create new habitat to replace that which is inevitably lost as a natural consequence of a changing climate? And can the areas that benefit from this water pay for the needed flood infrastructure elsewhere? This multifaceted problem requires a new sense of regional collaboration and the creation of a vision of a shared future.

As resource managers attend to this shared future, a harsh spotlight has fallen on the state of U.S. infrastructure. In the area of water and wastewater, the replacement and refurbishing estimates are in the trillions. Cities have come to realize that preventative maintenance and refurbishment have been greatly underfunded for most of the twentieth century and that the aging plumbing systems now struggling to accommodate to the needs of a growing population will fail under the added pressure of severe weather events. National focus and investment are needed to effect a remedy. But in any consideration of which financial tools should be used to help utilities, certain twentieth-century structures and practices need to be addressed.

Whether the urban utility is served by a public government utility or an investor-owned utility is immaterial. A notable portion of the water and sewer rates collected is not reinvested in the system. Those dollars either are paying shareholder profits or are redirected to fund other urban services. In cities where the government has chosen to privatize the system, the ratepayer has mostly paid for both, owing to municipalities' reliance on that revenue stream. Much is being said now about the "value" of water. Frustrations in the industry center on the reluctance of the ratepayer to fund even the smallest rate increase, with the result that utilities are unable to make long overdue critical improvements. The public has rightfully come to view water and sewer charges not as a utility bill, like power, gas, and cable bills, but rather as a tax that the public pays without a clear understanding of how or where that money is used.

In some parts of the United States, state law precludes that level of general government subsidy; elsewhere the realization has set in that to increase public willingness to fund such projects, government structures connected to the utilities must change. Newly conceived public corporations are emerging, as has happened in Louisville, Kentucky, where a new direct governance structure has been developed with one shareholder, the City of Louisville. In such areas utility budgets are no longer a subset of the city's general fund budget but are discrete and separate funds. Revenues are derived from the users, and expenditures are restricted to the benefit of the utility's ratepayers. In this setting, asset management and replacement can be approached from the longer-term perspective and ratepayers have a greater sense of transparency regarding how their dollars are spent.

In an examination of the larger issues that have emerged as the nation struggles to adapt to a changing climate and altered weather patterns, the regulatory structure cannot be ignored. Does it make sense that a desalination facility on the California coast should require ten years and untold millions to permit? Or that cities have found themselves at odds with the Clean Water Act as they have sought to incorporate green infrastructure into the redesign of their aging storm-sewer systems? Or that application of the Endangered Species Act has prevented the replacement of needed infrastructure while failing to save the species the act was intended to preserve?

Efforts to adapt to constant change and uncertainty are under way in an environment of regulatory rigidity that has emerged from the Safe

Drinking Water Act, the Clean Water Act, and the Endangered Species Act. These acts are administered by a patchwork of single-focus agencies with different missions. Since the passage of these congressional acts in the 1970s, they have existed and functioned in a sphere of relative certainty. The challenge for them today is to find a pathway to a more collaborative and flexible structure that would allow them to make course corrections as conditions changed. At the moment, most of what is being experienced is regulatory paralysis.

As the fallout from climate change raises questions about the function and provisions of federal, state, and local structures, it also challenges individual attitudes and behaviors. Americans take water for granted, and the notion that everyone shares a stewardship responsibility for the resource is only slowly emerging. Human consumption habits are a critical ingredient in finding a path to coexistence. We have lived by forcing our environment to adapt to our desired lifestyle and have given little thought to the notion that we must adapt to the surroundings in which we choose to live. Every living thing has a water footprint. In the twenty-first century it is the size of that footprint that matters. The water footprint consists not only of the amount of water used but also of the amount of waste and pollution returned by users. The passive customer of the twentieth century has to be transformed into the engaged customer whose self-awareness brings about critical behavior change. Otherwise it will be impossible to meet the water-related needs of a planet with 9 billion people on it.

That spring day in 2002 began the journey of adaptation for southern Nevada. The community today looks very different from how it did fifteen years ago. Where there was grass there is now desert landscaping. Where the average resident knew little about his or her own water use in relation to supply and took the resource for granted, there is now a far greater sense of stewardship and responsibility. Resource planning is firmly seen as a collaborative process between codependents of the water source, and infrastructure never imagined is being built, paid for entirely by the customers of the local utilities. This same journey is being undertaken in regions across the country, and regardless of the specific locale or the specific consequences of a shifting climate that are being experienced, similar challenges are emerging in structural, political, regulatory, and behavioral arenas.

Some areas of the world are ahead of the United States in managing and safeguarding their water resources. Both Israel and Singapore have developed national water resilience strategies and have invested and made the necessary structural changes to implement them. Australia has emerged from its devastating drought on the Murray Darling River with a dramatically altered legal framework and resource management strategies. Scotland has rethought its governance and utility structure, and China has put a secure water supply at the top of its national priorities. As Americans are jarred out of their complacency by the short- and long-term effects of increased flooding and sustained droughts, the need to face the reality of a crumbling and failing infrastructure and water security is finally rising to a position of priority on the national agenda.

Chapter 1

Climate Change
A Strategic Opportunity for Water Managers

KATHY JACOBS AND PAUL FLEMING

Water-related impacts associated with climate change are already evident in every part of the globe,[1] so water utility managers are faced with new challenges in providing the essential services on which the public relies. What types of skill sets and relationships must water resource organizations nurture to address the risks and opportunities associated with climate change? There is a strong tendency among water managers to rely on proven approaches and historical precedent because those approaches have worked in the past and are viewed as "safer" within the industry. But there are almost always multiple paths forward to address problems; some just require more persistence, innovation, and resources than others. The critical factor is to *embrace change* and expand adaptive capacity to prepare for change in building resilience as part of a professional code of conduct instead of relying solely on existing approaches.

The delivery of essential services while managing risks is the fundamental task of water resource managers, who have always worked in an uncertain environment. Their job is to make sure that water continues to flow through the taps in people's homes and businesses, and critical infrastructure remains intact, despite floods, fires, droughts, freezing temperatures, and heat waves. Water managers have always made decisions about the design and operations of water systems, reservoirs, and treatment plants without perfect information. They have not known how fast their community would grow, how much economic growth or contraction would occur in their area, or how much rain would fall in a given year. And they have always had to hedge their bets to protect public health and ensure a reliable water supply in the face of climate variability and numerous other sources of variability. Some of the sources of variability include changes that derive from the choices humans make, such as demographic shifts and changing economic patterns, as well as changes in natural systems, such as in the snowpack, ocean currents, and environmental water demands.

Historically, the tools for managing hydrological risk included reservoirs for flood control and water supply (to even out the peaks and troughs in precipitation and runoff), conservation (to limit peak or total demand), and a range of augmentation options, including importing water, the treatment and reuse of municipal wastewater, harvesting stormwater and storing it underground for future use, and desalination. The future, however, may require a broader set of water management tools, including more flexible institutional and regulatory arrangements. This is partly because of climate change, but also because many other factors are changing more rapidly than they did in the past. The combined effects of rapid changes in economies and land uses, in demographics and personal preferences about water use, and changing regulatory conditions and water rights issues, paint a picture of increasing sources of uncertainty even in the absence of climate change. The new reality is that climate change, which the U.S. Department of Defense has characterized as a "threat multiplier," will lead to unanticipated feedbacks and new and amplified sources of both risk and opportunity.

Although demands on the twenty-first-century water manager are in many ways similar to those that faced water managers in the twentieth century, climate change does bring a new array of challenges and opportunities. One way to get beyond traditional training and work toward a future that is more prepared for both is through partnering with other utilities, as well as through working with local universities and research networks. The opportunities that partnerships bring are dramatic and can be very useful in managing risk and encouraging innovation.

Adopting a Risk Management Framework

Managing risk is about managing perceptions as much as it is about managing the "real" risks that might be identified by experts. Though many Americans are not convinced that climate change is real or that it is caused by humans, that is not a good reason to ignore it. The failure to acknowledge the implications of climate change increases the risk of system failure for water managers. Especially for those who are operating in the "reality gap" (that is, working with customers or board members who are not willing to consider the scientific evidence of climate change), one approach is to frame decisions in the language of due diligence and risk management. For example, even if there is only a small chance that change climate is occurring, being wrong can have serious consequences. Properly applying a risk management framework means that all risks need to be considered, especially those with severe consequences, even if there is a low probability of a particular extreme event occurring. This explanation should provide sufficient space to permit consideration of climate change in the context of varied political perspectives. Being prepared for a wide range of future conditions is simply best practice for water managers.

Given that 97 percent of climate scientists agree that climate change is already occurring,[2] and that changes in the water cycle have been documented on every continent,[3] water utility managers are on the front line of the climate conversation. Acknowledging the human-caused component of climate change is important, particularly in the context of longer-term decisions. This is because the human influence alters the nature of the future we need to be prepared for. For example, if we were to assume that every change we are now observing is due to natural variability, then we would expect a return to normal or historical average conditions as we understand them. This has totally different implications for long-term water management than acknowledging that there is a driver of change (increasing total energy in the Earth's atmosphere) that is push-ing the system out of equilibrium and potentially outside the envelope of historical variability. Understanding the changes we are seeing through this second lens tells us that the impacts we are now observing are likely to escalate over time unless total global emissions actually decline in the near future, which seems unlikely.[4] Understanding that there is a trajec-tory of change rather than simple variability around a "normal condition"—in other words, understanding that stationarity is dead[5]—leads to a very different kind of approach for a water resource manager.

Water and Climate

The water cycle integrates social, environmental, and physical systems in multiple ways. Alterations in the water cycle are driving the impacts of cli-mate change through changes in snowpack, precipitation, flooding, season-ality of runoff and other events. For example, a reduction in total snowpack (due to more rain and less snow) and earlier snowmelt are having signifi-cant impacts on water-supply conditions across the U.S. West—in Califor-nia, the Pacific Northwest, and the Colorado River basin. More intense precipitation has been observed in every region of the United States,[6] even in places where the total rainfall is declining. More intense heat and drought conditions have also been linked to climate change in multiple parts of the country. These changes have a range of cascading effects on ecosystems, water supply, water quality, flood control, and economic sec-tors such as forestry, agriculture, and coastal and marine systems. Of course, there have been floods and droughts in the past, but the combination of the social, ecological, and physical changes that are occurring now means that the pace of change is accelerating and the challenges for water man-agers are also growing.

What does "stationarity is dead" mean in a water management con-text? It means changes in average conditions are occurring, but, more important, it means changes in extremes, which could have dramatic consequences. It means that historical conditions are no longer a reliable basis for future planning and that static standards, such as engineering rule curves for dam operations or intensity/duration/frequency curves for sizing drainage systems, are no longer optimal. It means more uncertainty, perhaps more variability, but also the opportunity to think broadly about water-supply reliability and the secondary implications for the full range of water resource management business practices, planning approaches, and decision-making processes, in the context of a range of other factors.

Climate is a driver of water-supply conditions, and changes in snowpack, temperature, and precipitation intensity, seasonality, or volume lead to a wide range of potential risks that ripple across economic sectors and communities. These changes affect demand for water, assets and infrastructure of utilities, and the capacity to provide reliable services, but they also have implications throughout and beyond the water sector. They affect energy supply and demand, agricultural productivity, and both natural and urban systems. The crosscutting, multifaceted implications of climate change mean that water resource managers will not only need to consider the associated technical and scientific issues but will also need to engage in understanding and managing the strategic, policy, political, and organizational implications of climate change across multiple sectors as well. It is imperative that water resource managers reimagine their roles and responsibilities in light of the challenges of a changing climate.

Institutions and Water Governance

Institutions, rules, regulations, and operating procedures (collectively known as the "decision context") can be a way to facilitate better outcomes. In many cases, however, the decision context is a serious constraint on available management options. The multiple layers of regulatory authority and intersecting and overlapping jurisdictions in the water world result in complex water governance in most regions of the United States. For example, there may be as many as 5,000 water management entities in the state of California.[7] There are advantages to some of this institutional complexity—institutions do provide stability and certainty in many cases—but complexity can also limit flexibility to respond in a rapidly changing world.

The long-lived implications of many water management decisions, such as constructing flood-control or water-supply infrastructure, also affect the options available to water managers. Though significant complexity in the legal, institutional, regulatory, and policy environments constrain the solution options that are available, there are important benefits to constraining the options. For example, water rights systems protect existing investments while limiting access of new users to water supplies. There are also some well-designed regulatory systems that explicitly constrain current options in order to provide future flexibility in times of severe shortages, such as the shortage-sharing criteria for managing the reservoirs on the Colorado River.[8]

The climate is not the only thing that is changing; the expectations of society are changing too. In some cases, changing societal values can be seen in the reallocation of water supplies and new challenges for water managers. For example, water that is now dedicated to tribal water settlements might previously have belonged to an irrigation district or a city. In Arizona, roughly half of the annual allocations of Central Arizona Project water have recently been dedicated to Indian water rights settlements.[9] Flow requirements to protect fish species in the San Francisco Bay–Sacramento-San Joaquin River Delta region have also affected California's water supplies in recent decades. Changing recreation patterns affect seasonal demands for water, including pressure for more in-stream flows in the summer. And changes in cropping patterns, for example more irrigation of agriculture in the Midwest and Southeast, are also affecting water-supply availability for the environment and other water users.

Further, water systems do not function in isolation from other systems. For example, they are strongly tied to and dependent on the energy grid. This is primarily because pumping water requires enormous amounts of energy (the California State Water Project is the largest single user of energy in the state) but also because most modes of generating electric power require substantial amounts of water. There are multiple implications of this interconnectedness, but in a risk management context it is most important to understand that there can be cascading effects of a failure in one part of a linked system. For example, if the energy grid or the communication system goes down in an extreme event, water availability may also be curtailed. Understanding the potential points of failure in the water system means understanding a much broader array of topics than has historically been the case and preparing for new kinds of risk, such as the risks that can cascade across interdependent "lifeline" sectors.

Implications of Politics and Public Perceptions

Even though at first glance a lot of the rules and regulations in place today do not make sense, there is almost always a historical rationale for their existence. Understanding which rules and regulations are truly obsolete and which are actually linchpins in a broader set of protections for the public or the environment requires very sophisticated analysis. These subtleties often are not well understood by stakeholders, who may gravitate toward solutions that are difficult to implement from a technical, legal, or economic perspective ("Why not build more dams to control storm runoff rather than wasting it?" "Why can't we do more with desalination?"). Some solutions that seem obvious to water management professionals are likewise nonsensical to the public even if they are technically feasible, for example, toilet-to-tap approaches to reusing municipal wastewater, or investments that might "strand assets" for a significant period of time while also providing a range of other benefits in the near term.

If climate change is seen as a lens through which to view the future, it allows the reframing of issues that have needed to be reconsidered for a long time. This reframing can help water managers and other decision-makers who historically have been isolated from broader policy discussions to enter the conversation for the first time. For example, a broad risk management strategy in a climate change context almost always requires discussion of land use, energy supply, public health, and environmental quality. The highly complex and integrated nature of climate change impacts means that discussions about solutions require an interdisciplinary and wide-ranging set of considerations. These broader conversations have led to very different kinds of planning for the future in places like New York, where "asking the climate question" has led to a much more integrated approach to planning across agencies and disciplines. The disaster resilience community has also embraced a broader view of preparation for and recovery from extreme events. For example, the Housing and Urban Development Disaster Resilience Competition,[10] required an interdisciplinary, intersectoral approach to planning. In activities related to climate adaptation promoted by the National Oceanic and Atmospheric Administration (NOAA) and Department of the Interior (DOI) federal science networks,[11] there is evidence that climate change discussions open doors to new ideas and more inclusive approaches. Seattle's experience provides good evidence that planning for the long term also has short-term benefits, in part because the climate impacts projected long into the future are arriving more quickly than expected (see box 1-1).

New Solutions in a Changing World: The Role of Partnerships

Tactics to enhance understanding and to assess the nature of climate change and its impacts on the water cycle and water resources include establishing collaborative relationships with the climate research community and strengthening peer-to-peer learning networks between utilities. Building relationships with the climate research community is one useful way to gain access to the best available science regarding how climate change may play out in the geographic area where a utility operates, without needing to employ a whole team of climate modelers. These relationships can and should allow for two-way conversations, serving as a means for researchers to disseminate relevant research and reports to utility managers but also allowing a reality check on that information as utility managers provide feedback on the topics and the usefulness of the outcomes. Funding university researchers to conduct original research to address specific information needs identified by the utility manager is one way to build these relationships. Investments in working with local research universities can reap greater rewards if the relationship between the utility and the researcher is maintained over time, enhancing both trust in the results and the ability to do longitudinal studies. There are also numerous cases in which utilities have collectively funded researchers

BOX 1-1

Seattle: Managing Changing Hydrological Conditions

For Seattle Public Utilities (SPU), the winter of 2015 foreshadowed what climate conditions might look like in the future. When observed temperatures in 2015 were compared with a range of projected temperatures for 2050 derived from downscaled climate models, the temperature in the winter of 2015 was at the warmer end of the projected temperatures; observed precipitation levels were also at the drier end of the range of climate projections. The exceptionally warm temperatures led to record low levels of snowpack. Recognizing the implications of these conditions, SPU shifted its operations from flood control to reservoir refill earlier than it typically would have. Dynamic reservoir management is an operational adaptation strategy that may have to be increasingly deployed by water managers as they strive to meet their multiple responsibilities and objectives in light of a changing hydrograph driven by climate change. While dynamic management can be a critical component of an adaptation strategy, utilities will most likely need to evaluate a full suite of adaptation options to develop a robust water management strategy in the face of climate change.

to respond to questions that have relevance to many parties, for example, through the Water Utility Climate Alliance.

When utilities and researchers decide to enter into cooperative agreements to pursue research, it is critical that both the utility and the researchers be clear about their own objectives and about the needs and expectations of both parties. Historically in such endeavors, many unexpected challenges have arisen because of a failure to establish clear expectations. For example, while university-based researchers must publish their work in journals, utility managers may prefer not to have their climate-related risks and vulnerabilities characterized by researchers in journal articles, especially if they are not part of the author team. Most water managers are not focused on publishing a peer-reviewed article as an outcome of their work. Rather, they are typically looking for a technical report pertinent to their information needs, and they may or may not support releasing the information publicly. In effect, water utility managers and researchers need to determine whether their research relationship is a consultant-client relationship, which utilities are most accustomed to, or a funder-researcher relationship, which is probably what researchers are most accustomed to. Clearly establishing the basis of a research arrangement and the various roles and responsibilities before the research begins will help manage expectations and avoid potential conflict.

Increasingly, the relationship between utilities and researchers is being viewed not as a one-way flow of information but as a two-way dialogue based on the coproduction of knowledge. With this framing, the applied and tacit knowledge that utilities have of their own systems is coupled with the deep science-based knowledge researchers have of projected climate change to coproduce knowledge that leverages the best of both kinds of information. Coproduction relies on joint learning, which usually means focusing research on the applied information needs of the utility so that the user of climate research is also the shaper of that research.[12] This requires both utilities and researchers to rethink their roles in the scientific enterprise. The utilities' role requires that they become adept at identifying and articulating their needs and be able to disclose and frame their system knowledge in the context of climate change as well as be able to conduct some of the analysis itself. Meanwhile, researchers must be responsive to the information needs of utilities, be willing to deviate from a predetermined research agenda, and learn how to translate their technical findings in a manner that resonates with the utility. Robust and meaningful utility engagement in coproduction may require that the utility conduct an honest assessment of its capacity to sustain this type of engagement over time and a determination of whether it has the personnel with the right skills to lead this effort.

Generating climate information based on user needs and developing and sharing findings with decision-makers is an example of how to provide climate services in a co-production mode. While efforts have been made to establish a national climate service for the United States, those efforts have not borne fruit for a number of reasons, including politics and economics. Short of a full-blown climate service, which seems unlikely to materialize in the near future, there are numerous federal and university-based research initiatives that water utilities can tap into to enhance their understanding of climate impacts. NOAA's Regional Integrated Science Assessments (RISA) program and the DOI's Climate Science Centers (funded through the U.S. Geological Survey) are two of the leading applied federal climate research programs. Several other federal databases and tools exist, many of which are included in the U.S. Climate Resilience Toolkit created by the White House Office of Science and Technology Policy.[13] A particularly relevant example is the U.S. Environmental Protection Agency's CREAT tool, which assists water utilities in understanding climate impacts on their systems.

Collaborative relationships with the research community are only one approach to accessing climate information to support risk management efforts. Strengthening peer-to-peer learning and collaborative research projects with other utilities can open other paths to innovation. Sustained peer-to-peer learning provides an unrivaled opportunity to enhance understanding of climate-related challenges and develop approaches for managing them. It is clear that the trusted relationships that exist within professional societies are an excellent means of conveying information in ways that are useful, timely, and credible to water managers. Often, these approaches provide the professional "safe space" in which to explore new ideas without taking the personal risk often associated with unusual or untested ideas.

The Social and Economic Components of Climate Impacts: An Opportunity for Reframing

Utilities need to be proactive in preparing for future changes and to manage risk. There are multiple components to anticipating future conditions and being resilient that extend well beyond projections of future water-supply conditions. They include robust communication strategies (internal and external); infrastructure investments; understanding the demand side of the equation (how climate may affect demand for water, for example); and the potential for behavior modifications, such as conservation; and robust approaches to strategic planning that consider a broad spectrum of plausible futures. Utilities must also consider climate impacts through a service equity lens. The disproportionate impacts that climate change will have on different segments of society need to be understood and strategies to address those future impacts—and current inequities—need to be embedded into utility practices so that the essential services on which communities rely are equitable for all.

Understanding how climate change will alter the water cycle and how those alterations will physically affect supplies and resources is now an essential responsibility of an effective water utility. It is increasingly clear, however, that a water utility must also be comprehensively evaluating how climate change will affect its core functions—from strategy to financing to communications—to fully understand the implications of climate change and to be in the best position to manage the associated risks and leverage opportunities. In short, climate change should not be viewed as purely an environmental, technical, or water-supply challenge but as an issue that has crosscutting implications that must be understood by utility managers.

The intersection of water management and climate change brings additional considerations related to the role of water in supporting a wide range of economic sectors and activities. Climate change affects agriculture, electric power, forestry, transportation, industries, and recreational choices in a multitude of ways that may need to be considered when thinking about future water-supply needs, the potentially increasing competition for limited supplies, and changing demand patterns. It is now widely recognized that exported products contain "embedded water"; for example, grain or alfalfa that is grown in irrigated fields for export requires a certain number of gallons of water to be produced, processed, and shipped to a particular location. The globalization of the economy means that climate impacts that occur in one part of the world can affect production and product availability in other parts of the world. All of these considerations affect water management in one way or another.

One important way that perceptions around water and climate change are evolving is in how the risks of water crises and climate change are being characterized. For example, each year the World Economic Forum (WEF) issues an annual report documenting the top global risks from the perspective of 800-plus business, academic, government, and NGO leaders.

In 2015, water crises and the failure of climate adaptation were two of the greatest risks in terms of impacts and likelihood, with water crises being viewed as a societal risk rather than an environmental risk, as it was in the past. With the strong linkage between water crises and climate change, the inclusion of both these risks at the top of the list is telling.

The Risky Business Project, which documented the economic impacts of climate change in three different economic sectors in the United States, was instrumental in taking the extensive understanding of the physical impacts of climate change and translating it into economic impacts.[14] Although this project was necessarily limited in scope, the magnitude of the economic impacts of climate change within regions was dramatic. These findings, along with the WEF's framing of water crises as a societal risk, is an invitation for water managers to translate the myriad water-cycle impacts of climate change into something more meaningful for their constituents and board members: the potential economic costs to society of climate disruption of the water supply.

In a third recent example, the *Harvard Business Review*, in its April 2014 cover story, argued that being a resilient organization in the face of climate change requires rethinking corporate strategy.[15] The story refers to a big pivot in strategic thinking that emphasizes collaboration across the competitive landscape, the use of science to inform decisionmaking, and changes in valuation techniques. This reframing of climate change as a driver of fundamental repositioning rather than as just an environmental issue is echoed in a previous WEF Global Risk report from 2013. In that report, the term "climate smart mindset" was used to refer to the incorporation of climate considerations from the strategic level of an organization to the operational level.[16]

In light of this strategic and comprehensive framing of climate change in some quarters of the private sector, it is perhaps not surprising that climate change issues are now cropping up in the financial sector in ways that are of potential relevance to utilities. Some examples are the fossil fuel divestiture movement and the continued data gathering, surveying, and analytical activities of groups like the Carbon Disclosure Project. But closer to home for water utilities is the emergence of a significant increase in activity in the green bond or climate bond arena. HSBC, a large financial services firm, projected that the issuances of green or climate bonds in 2016 would grow between 32 percent and 91 percent over 2015 levels.[17] Though such bonds are still only a small sliver of the total bond market, growing investor appetite for green or climate bonds indicates the emergence of a niche in which climate change and finance come together. In aggregate, these examples illustrate how climate change is increasingly being viewed through multiple lenses, from strategy to finance to operations, perspectives that a forward-looking water utility must be considering in addition to the physical impacts of climate change.

Climate Communications: More Opportunities for Reframing

Another surprising area of opportunity is communicating climate change impacts to the public. The conventional wisdom is that climate change is an issue to be avoided when engaging with the public. For example, in the 2012 presidential election, climate change barely surfaced as an issue because strategists advised against it. While polling results on the issue certainly vary, polling in early 2015 indicated that this conventional view may be shortsighted. A *New York Times*, Stanford University, and Resources for the Future poll from late January 2015 noted that an overwhelming majority of the American public, including half of Republicans, support government action to curb global warming, and that the number of people who believe climate change is caused by human activity is growing.[18] The same three organizations examined Hispanic views on climate change and found that 54 percent of Hispanics polled listed global warming as extremely or very important to them personally and 67 percent felt they would be hurt personally if nothing was done to reduce global warming.[19] According to a Clinton Global Initiative and Microsoft poll, 66 percent of millennials say there is solid evidence the Earth is getting warmer and 75 percent say human activity is responsible for it.[20] And with respect to the last two polling examples, there is the potential to embed climate considerations into a water utility's hiring strategies by including a proactive stance on climate change in the recruitment of millennial and Hispanic populations. Owing to the utility sector's current issues with an aging workforce, employee retention and recruitment will likely take on greater importance as utilities grapple with succession planning.

The most startling poll results are from a study funded by the Water Research Foundation. Led by Stratus Consulting, the project worked with Tony Leiserowitz of the Yale Project on Climate Communications to survey 1,000 people on their views of climate change and water utilities. Some of the results were: (1) 73 percent of the respondents agreed with the statement that climate change will have a significant impact on the water cycle; (2) water utilities were viewed as trustworthy sources of information on climate change impacts on local water systems by 71 percent of respondents, more favorably than environmental groups were viewed and nearly on par with local or state colleges and universities; and (3) 92 percent of respondents wanted water utilities to play a leadership role in helping their community prepare for the impacts of climate change.[21] In short, water utilities have significant reputational capital on the issue of climate change, and that capital is most likely largely untapped and underutilized.

All of these developments—the consideration of climate change as it relates to strategic, financial, and operational issues, the growing support for climate action in the general populace, the extensive trust and support that the general public has for water utilities on climate change—are playing out in the context of an aging, deteriorating, and underfunded water infrastructure in the United States. This brings us back to the argument that utilities need to engage with, at least initially: understanding how climate change will affect the water cycle, its impacts on the assets and infrastructure utilities manage, and its effects on the essential services they deliver. It is now an imperative for any forward-looking water utility to understand the nature of these climate effects now and as it plans to replace aging infrastructure and address investment backlogs. But utilities also need to move beyond the physical implications of climate change for their assets and services to explore the implications for other, related aspects of utility decisionmaking and business practices. Understanding how to leverage utility reputational capital and public opinions on climate change to facilitate support for those investments and strategically reframe climate change as an issue that affects all of a utility's core functions is an opportunity not to be missed.

Conclusion

Historically, water management decisions have often been made in reaction to extreme weather events or based on historical patterns and statistics. For a wide array of reasons, proactive consideration of the implications of climate change presents new opportunities to make investments in water management that will provide multiple benefits to water utilities and their customers. The opportunities discussed here include the positive benefits of an introspective utility-wide look at "whole-system" (socioeconomic, physical, ecological) health in the context of communities and ecosystems, the potential for partnerships and peer-to-peer learning to encourage innovation, and new ways to approach climate change impacts in terms of both managing risks and seizing opportunities. Finally, it is clear from recent work in climate communications and the obvious linkages between water supply and the economic health of communities and ecosystems that water utilities are well positioned to provide leadership on climate resilience issues.

To seize the opportunities and to avoid as many surprises as possible, it is important that utilities proactively start evolving and reframing their roles and responsibilities in light of climate change. Fundamentally, this involves building adaptive capacity, either through drawing on staff knowledge or fostering new areas of expertise, activating professional knowledge networks (peer-to-peer, researcher-to-practitioner, and so forth), or developing long-term relationships with local research universities. This explicit and sustained commitment to reframing the roles and responsibilities of a utility is likely to help address the various challenges and opportunities posed by climate change, as well as make the communities served by these utilities more resilient to broader societal issues as well.

Notes

1. *Working Group I Contribution to the Fifth Assessment Report of the International Panel on Climate Change 2013: The Physical Science Basis* (IPCC, 2013); Jerry M. Melillo, Terese (T. C.) Richmond, and Gary W. Yohe, eds., *Climate Change Impacts in the United States: The Third National Climate Assessment*, U.S. Global Change Research Program, 841 (2014) (doi:10.7930/J0Z31WJ2).

2. E. Maibach, T. Myers, and A. Leiserowitz, "Climate Scientists Need to Set the Record Straight: There Is a Scientific Consensus That Human-Caused Climate Change Is Happening," *Earth's Future* 2 (2014), pp. 295–98.

3. *Working Group I Contribution to the Fifth Assessment Report of the International Panel on Climate Change 2013: The Physical Science Basis* (IPCC, 2013); Jerry M. Melillo, Terese (T. C.) Richmond, and Gary W. Yohe, eds., *Climate Change Impacts in the United States: The Third National Climate Assessment*, U.S. Global Change Research Program, 841 (2014) (doi:10.7930/J0Z31WJ2).

4. P. C. D. Milly, Julio Betancourt, Malin Falkenmark, Robert M. Hirsch, Zbigniew W. Kundzewicz, Dennis P. Lettenmaier, and Ronald J. Stouffer, "Stationarity Is Dead: Whither Water Management," *Science* 319, no. 5863 (February 1, 2008), pp. 573–74.

5. Ibid.

6. A. Georgakakos, P. Fleming, M. Dettinger, C. Peters-Lidard, Terese (T. C.) Richmond, K. Reckhow, K. White, and D. Yates, "Water Resources," in *Climate Change Impacts in the United States: The Third National Climate Assessment*, edited by J. M. Melillo, Terese (T. C.) Richmond, and G. W. Yohe, U.S. Global Change Research Program, 69–112 (2014) (doi:10.7930/J0G44N6T).

7. Lester Snow, telephone communication with Jacobs, 2015.

8. U.S. Department of the Interior, Bureau of Reclamation, "Colorado River Interim Guidelines for Lower Basin Shortages and Coordinated Operations for Lake Powell and Lake Mead," 2007 (http://www.usbr.gov/lc/region/programs/strategies.html).

9. Rosalind H. Bark and Katharine L. Jacobs, "Indian Water Rights Settlements and Water Management Innovations: The Role of the Arizona Water Settlements Act," *Water Resources Research* 45, no. 5 (May 2009), art. W05417 (doi: 10.1029/2008WR007130).

10. Federal Register /Vol. 81, No. 109 /Tuesday, June 7, 2016 /Notices: Department of Housing and Urban Development [Docket No. FR–5936–N–01], Notice of National Disaster Resilience Competition Grant Requirements, Office of the Assistant Secretary for Community Planning and Development, HUD (http://www.lexissecuritiesmosaic.com/gateway/fedreg/2016-13430.pdf).

11. Radley Horton, Cynthia Rosenzweig, William Solecki, Daniel Bader, and Linda Sohl, "Climate Science for Decision-Making in the New York Metropolitan Region," in *Climate in Context: Science and Society Partnering for Adaptation*, edited by Adam S. Parris, Gregg M. Garfin, Kirstin Dow, Ryan Meyer, and Sarah L. Close (New York: Wiley, 2015), pp. 51–72.

12. A. Meadow, D. Ferguson, Z. Guido, A. Horangic, G. Owen, and T. Wall. "Moving Toward the Deliberate Co-Production of Climate Science Knowledge," *Weather, Climate, and Society* 7, no. 2 (2015), pp. 179–91 (doi: http://dx.doi.org/10.1175/WCAS-D-14-00050.1).

13. U.S. Climate Resilience Toolkit (https://toolkit.climate.gov/), developed by the Obama administration.

14. Risky Business, *The Economic Risks of Climate Change in the United States* (June 2014, updated September 8, 2014) (http://riskybusiness.org/site/assets /uploads/2015/09/RiskyBusiness_Report_WEB_09_08_14.pdf). The Risky Business Project was cochaired by Michael R. Bloomberg, former mayor of New York City, Henry M. Paulson Jr., former secretary of the treasury, and Thomas F. Steyer, founder of Farallon Capital Management.

15. Andrew Winston, "Resilience in a Hotter World," *Harvard Business Review,* April 2014 (https://hbr.org/2014/04/resilience-in-a-hotter-world).

16. World Economic Forum, *Global Risks 2013 Eighth Edition* (http://www3 .weforum.org/docs/WEF_GlobalRisks_Report_2013.pdf).

17. Jessica Shankleman, "Green Bond Market Will Grow to $158 Billion in 2016, HSBC Says," Bloomberg, January 26, 2016 (www.bloomberg.com/news/articles/2016 -01-26/green-bond-market-will-grow-to-158-billion-in-2016-hsbc-says).

18. Coral Davenport and Marjorie Connelly, "Most Republicans Say They Back Climate Action, Poll Finds," *New York Times,* January 30, 2015 (http://www .nytimes.com/2015/01/31/us/politics/most-americans-support-government-action -on-climate-change-poll-finds.html).

19. Coral Davenport, "Climate Is Big Issue for Hispanics, and Personal," *New York Times,* February 9, 2015 (http://www.nytimes.com/2015/02/10/us/politics /climate-change-is-of-growing-personal-concern-to-us-hispanics-poll-finds.html).

20. C. Jane Timm, "Millenials: We Care More about the Environment," MSNBC .com, March 22, 2014 (http://www.msnbc.com/morning-joe/millennials-environment -climate-change).

21. R. Raucher, K. Raucher, A. Leiserowitz, S. Conrad, M. Millan, B. Dugan, and E. Horsch, *Effective Climate Change Communication for Water Utilities* (Denver, CO: Water Research Foundation, 2014).

Chapter 2

The Sacramento-San Joaquin River Delta
Resolving California's Water Conundrum

PAT MULROY

On a hot summer morning in the Sacramento-San Joaquin River Delta, California, researcher Jon Burau of the U.S. Geological Survey (USGS) and his aluminum-hulled monitoring boat are roaring up the main stem of the Sacramento River. From the bow, the delta is a meandering, monotonous blur of muddy green water bordered by gray rocks along the earthen levees. Above is a cloudless blue sky. Heading north from the city of Rio Vista toward Sacramento, Burau makes a sudden right turn at the entrance of a small inlet known as Miner Slough. Here he turns off the engines and lets the boat glide nose-first into a thick bed of green tules and young willow trees. The 250-horsepower outboard motor chortles to a halt. Suddenly the delta is still and silent. It is an opportunity for Burau to briefly convert from captain to professor and discuss what ails the surroundings at Miner Slough.

A few years ago, this spot was like many others in the delta. It was bordered by rock and devoid of fish-friendly vegetation. But as a cost-cutting measure, a portion of the levee was set back so that the daily tides could once again roll over the land, as they once did throughout the entire delta. A functioning delta is an ever-changing interface between water and land.

It is a labyrinth of marshes filling and draining and breathing life with each tidal cycle. Little about today's delta is functioning as nature—and now governments—intended except on this small section of river on Miner Slough and a very few other places in the modern delta.

The West Coast's largest salmon run once passed through here. The young smolts, preparing for their ocean life, would hide, eat, and gain size and strength in tens of thousands of acres of marshlands. Now all but one salmon run that pass through this estuary are endangered. And when the smolts do migrate, they face a gauntlet of deep channels, rock levees offering no places to hide—except here in this postage stamp of habitat on Miner Slough. In terms of salmon-friendly delta habitat, says Burau before starting up the engine again on that hot August day in 2014 (with follow-up questions thereafter), "of the 1,110 miles of levees this is the only spot I know. We don't have any of these habitats left."

The delta is the epicenter of California water. The rivers of the western Sierra Nevada converge here before heading west to San Francisco Bay and the Pacific Ocean, forming the largest estuary on the West Coast. Few societies in the world have placed their water systems in the heart of an estuary. Yet here in California, the nation's two largest water-pumping facilities are located in the Delta. One of every six acres of irrigated farmland in the United States is sustained by this watershed, and economies from Silicon Valley to San Diego depend on it. The size and the stakes speak for themselves.

Since the 1990s, wildlife agencies have listed one dwindling fish population after another under the Endangered Species Act (ESA). They have largely tried to solve the problem by restricting the water systems, not by restoring habitat. A primary water supply for California has steadily decreased in both amount and reliability. When flows are at their highest, the pumping levels can be at their lowest—all because of the restrictions.

This severe regulatory approach has not been accompanied by the recovery of a single fish species. The once abundant pelagic organisms, such as the endangered delta smelt, continue to decline. Climate change appears to be playing an increasing role, particularly in the rising delta water temperatures, which are believed to contribute to the smelt's precipitous decline. Many recent monitoring efforts to find a single smelt have come up empty.

The debate over why rages on.

Habitat Conservation under a Regulatory Regime

The desperate circumstances in the delta have led to a cycle of ever-greater pumping restrictions and litigation. The result is nothing tangibly positive either for the San Francisco Bay or the Sacramento-San Joaquin River Delta ecosystem or for California's economy, which depends on water supplies that must pass through the delta.

"Once a fish becomes endangered, it becomes extremely difficult to study," says Burau, who nonetheless is trying to research the movements of salmon and smelt in studies throughout the estuary. "It is a catch 22." Studying a species means capturing some, taking some with what is called an "incidental take permit." "The permits you need to obtain a reasonable sample size are virtually impossible to get. And once fish are 'endangered,' they are hard to study because they are rare and hard to find, yet when a fish is placed on the endangered species list that is exactly when monies become available to study them, often after it is too late."

The last governor to propose a sweeping solution to the intractable problems afflicting the delta was a young upstart named Jerry Brown. He did so in 1980 before the problem had reached the stage of a full-blown crisis. It was a purely political approach. Brown and California lawmakers passed legislation to move the intake for the State Water Project thirty-some miles to the north on the Sacramento River and out of the heart of the delta. They proposed to transport the supply to the aqueducts in the southern delta by constructing a canal around the estuary, the so-called Peripheral Canal. And to protect in-delta and upstream water rights, the water-supply improvements were linked to various new legal and environmental protections.

In California, voters can overturn any act of the legislature. All it takes is enough valid signatures to overturn statutes or policy. And in 1980, that is what began to happen. One of the stranger sets of California political bedfellows forged an alliance. On one side of the coalition were environmental groups and northern interests who feared the Peripheral Canal would send too much water to Southern California and San Joaquin Valley farmers. They were joined by traditional adversaries, certain San Joaquin farmers who held precisely the opposite fear, that the plan had too many environmental protections and too little water.

Sufficient signatures were gathered to place the delta water legislation, California Senate Bill 200, on the November 1982 ballot. Tepid voter support in Southern California was overwhelmed by many northern counties voting more than 90 percent in opposition. Never before in an election had California been so regionally divided. Governor Brown had two more years in office, but he was done with the delta. Nothing approaching a comprehensive solution would take form for another generation.

By 2005 there was little choice other than for the state and federal governments to rethink the delta. The traditional method of enforcing the ESA, one fish species and one set of pumping restrictions at a time, was not working. Something had to be done. While there was nothing approaching a consensus on the solution, just about every stakeholder agreed that the status quo was not acceptable.

The provisions of the ESA seemed to offer one potential way out: regulatory oversight moving from the ESA's section 7 to section 10, the location of the so-called habitat conservation plan. The basic idea seemed straightforward. The applicants (in this case, the operators of the delta's State Water Project and federal Central Valley Project) would seek long-term coverage of their activities—that is, providing public water supplies—through new permits. In exchange, they and state and federal agencies would agree to a plan of long-term conservation actions (a modernized water system with fewer fish conflicts, plus restoration). A habitat conservation plan seemingly offered the prospect of fifty years of regulatory stability and a far more comprehensive set of conservation actions than the wildlife agencies could otherwise require.

Habitat conservation plans are commonplace throughout the United States. A quite successful one is under way on the Colorado River, the Lower Colorado River Multispecies Conservation Plan. In California, these plans protect birds such as marbled murrelets in the North Coast redwoods and gnatcatchers inhabiting undeveloped canyons in urban Southern California. Yet a habitat conservation plan in the United States has never been implemented anywhere for an aquatic environment like the delta, one of the most physically altered and politically tumultuous landscapes on Earth. The ESA is being put to its toughest test in one of the most difficult places imaginable.

Efforts to develop a habitat conservation plan for the delta began ten years and two governors ago in October 2006 in the form of the Bay Delta Conservation Plan (BDCP). And now in his fourth term and approaching his eighties, Governor Jerry Brown is back. He inherits water business left unfinished in the 1980s by himself and subsequently by Arnold Schwarzenegger. This time the solution is not being driven through the political process of a legislature dictating a given solution. Now Brown is seeking a regulatory solution. In western water, this is the mother of all permitting processes. It is a grand test of government agencies' ability to take action.

"Analysis paralysis is not why I came back thirty years later to handle some of the same issues," Brown told a press gathering in July 2012 when unveiling the framework of his delta strategy. "At this stage, as I see many of my friends dying—I went to the funeral of my best friend a couple of weeks ago—I want to get s _ _ _ done."[1]

As in 1980, the emerging plan calls for diverting some of the state's water supplies from the northern delta to minimize fish conflicts. And as in 1980, the plan calls for a new conveyance system. This time the engineering solution is three new intakes on the Sacramento River. They would feed thirty-four miles of twin tunnel pipelines that would transport the water to the existing aqueduct system. The tunnels would thereby avoid numerous drainage conflicts (and angry locals) that exist on the land above the tunnels.

What is new this time around, and what gives researchers like Burau a ray of hope, is a government desire for wide-scale habitat restoration. For young salmon trying to migrate through the delta to sea, says Burau, "there is no other way to help them—we can't fix this problem simply with water alone. At times the water is not available, like now, in a drought. Habitat, once created, helps organisms survive no matter how much water is available in a given year."

Water management in California has also dramatically changed since the 1980s, and for the better. Groundwater management is now the law of the land, thanks to a landmark 2014 statute. And in the urban world, conservation and local supplies have reduced reliance on the delta. Southern California has grown by 5 million people in the past generation while reducing its use of imported supplies—and plans to do the same for the next generation. The strategy for the delta has been dubbed "Big Gulp, Little Sip": to capture water when nature makes it available and take far less when it is dry.

Yet old fears die hard. As a delta solution has been slowly forming in recent years, a new set of strange bedfellows has been assembling in opposition to it. There is, for example, a somewhat traditional group of in-delta interests. It calls itself Restore the Delta. It is a curious name insofar as, based on the group's website content, it seems to be opposed to wide-scale delta restoration. Gary Kremen, the founder of Match.com and a resident of Palo Alto, has voiced opposition to the plan. He spent more than $300,000 of his own money to get elected to Silicon Valley's largest water district, a key player in the effort. Certain environmental groups, such as the Natural Resources Defense Council, try to straddle the fence, saying at times they have somewhat supported the process yet never putting a solution on the table. Others, such as the Sierra Club, attempt no nuance and are outright opposed.

California is fast approaching a decisive chapter with its delta. From the outside, this feels like one last valiant try, the most sophisticated and sweeping set of actions ever proposed, to solve this problem. Otherwise nature stands ready to reorder the landscape in one way or the other.

The Watershed and Water Rights

The watershed of the delta spans the western Sierra Nevada, some 27 percent of the state's land mass. The Sierra should be a source of unity for the state. Nearly 96 out of every 100 Californians rely on these mountains for some or all of their water. Yet the delta crisis makes the Sierra a source of deep statewide division.

The water agencies and interests representing those 96 percent of Californians break down into numerous factions. There are the senior water rights holders, whose interests predate the construction of dams. There are "independents," such as San Francisco, with its own dam-and-conveyance system, which taps the Tuolumne River high in the Sierra in Yosemite National Park. Another independent is the East Bay, with its sister system on the Mokelumne River. Contra Costa County maintains its own pumping system in the Delta, feeding its nearby reservoir, Los Vaqueros. And then there are "the contractors," communities from Silicon Valley to Bakersfield to San Diego that are partially dependent on appropriative water rights, based on reservoir construction. On the east side of the San Joaquin Valley the San Joaquin River, a tributary to the delta, has its own reservoir and canal system that provides water locally unless, in times of shortage, senior rights holders downstream get their water first.

California's complex water rights system is hardly the cause of the delta's problems, but it drives the water geopolitics of California. One stakeholder's solution can be quickly perceived as another's new problem. And with a state this big, and a water system so vast and complex, there has never been a delta solution that satisfies everyone.

The Colorado River serves seven western states and the Republic of Mexico. Approximately 14 million acre-feet of water are spoken for annually under the "Law of the River," a set of agreements and compacts that has arisen over time. The delta basin dwarfs the Colorado when it comes to precipitation. The western Sierra provides 32 million acre-feet of river flow in an average year. In a wet year, more than 30 million acre-feet of water can rush out to sea.

Overall, however, the human uses of Colorado River and Sacramento-San Joaquin River Delta waters are of surprisingly comparable quantities. That is because in California, roughly half the flow of the delta watershed makes it through the estuary and under the Golden Gate Bridge to the Pacific Ocean in an average year. By contrast, the highly regulated flow of the Colorado is entirely spoken for by one user or another.

Southern California, representing half the state's population, uses only 4 percent of the overall Sierra supply by way of the delta's State Water Project; the San Francisco Bay area uses another 3 percent; and farms and communities from Redding to Bakersfield use 41 percent. California outwardly appears to have a north-south dynamic when it comes to water. But it can be east-west. Farmers within the delta divert as much water as does the Metropolitan Water District of Southern California.

In the delta, the pumping facilities of the federal Central Valley Project have been providing supplies to the San Joaquin Valley and portions of the Bay Area since the 1940s. Governor Pat Brown led the effort (and a close statewide vote) to approve the California State Water Project in 1960. Operations began in the 1970s. Together, the Central Valley Project and the State Water Project export about 18 percent of the Sierra outflow in an average year. Upstream uses divert double that amount. The two large pumping projects are conspicuous targets for regulation in an estuary that is in trouble. They represent the two largest "knobs" that can be turned. But to the extent that diversions are contributing to the demise of the delta, any given use must be seen in a larger context. These facilities have increased diversions over the past century. But upstream diversions have also increased, and by nearly identical quantities, according to an analysis by the Delta Blue Ribbon Task Force. This watershed has many, many dependents. And for the delta, there are many other contributors to its decline.

Occupying the Marshlands: Restoring a Degraded Habitat

California, in a series of actions, essentially decided to occupy the marshland of the delta soon after the gold rush. The state has been living with the consequences of this and countless other decisions ever since.

Today's confected "islands" in the delta, with names like Bacon, Sherman, Twitchell, and Victoria, are entirely human creations. The construction of levees permanently severed the natural connection between tides and land. According to a 2014 study by the San Francisco Estuary Institute, the original delta had more than 3,217 kilometers of small channels less than fifteen inches wide, with perhaps another 1,900 kilometers un-mapped.[2] Today's delta has 144 kilometers of small channels. Those ha-vens for salmon, like that postage stamp at Miner Slough, are all artifacts of history.

The delta has more miles of man-made levees than California has miles of natural coastline. "The delta is basically a network of canals," says Burau. "They are deep. They are narrow. They are covered with rock. There is no variability in the current. No back eddies. It all just goes downstream." This is not by accident. The gold rush era, which included a devastating pattern of hydraulic mining in the Sierra before courts finally shut down the industry, left a legacy of sediment in the rivers downstream. The engi-neers who built the levees needed help to get rid of it; hence the deep chan-nels. "This configuration of the channels was intended, in part, to get the hydraulic mining debris out of the delta, primarily for flood protection," says Burau.

For today's young salmon trying to migrate downstream, the delta can be a graveyard. Monitoring experiments upstream on the San Joaquin River have found a predation rate as high as 100 percent. "The water just hauls," Burau says. "There are no velocity refugia in the system. There are no tules. The little guys [salmon] are in a situation where the predators have a big advantage."

The alterations go on. An estimated 95 percent of the delta watershed's original floodplain has been erased by levees and human developments, everything from the capital of Sacramento to the rice fields of the north-ern valley. Young salmon migrating downstream once found food and shelter on the floodplains, even those that drained quickly after a storm. "They are good at getting on and off these floodplains," says Burau. "It's not surprising—salmon evolved in these environments."

The original delta had 38,000 acres of riparian forest. It was a haven for migrating birds along the Pacific Flyway. Today's delta, according to the San Francisco Estuary Institute, has closer to 10,000 acres.[3]

The 2014 Estuary Institute report identified six drivers of ecological change in the Delta: an overall reduction in habitat, a loss of habitat di-versity, loss of habitat connectivity, degradation of habitat quality, lack of habitats that offer resilience, and the presence of non-native species.[4] None of these drivers of change to date has been a primary target of regulatory enforcement of the ESA. Pumping restrictions address none of these fundamental problems.

If there is one fish that illustrates the act's consequences for both scientists and water operators, it is the delta smelt.

The Smelt Story

A few miles downriver from Jon Burau's favorite salmon spot at Miner Slough, past the confluence of the Sacramento and San Joaquin Rivers, the delta becomes an expansive water superhighway. Burau's USGS boat feels like a blip on a vast body of water. By all appearances we are motoring on an inland sea. The incoming tides reverse flows from westward to eastward twice every day, moving water and organisms nine to ten miles twice each and every day. The valley's hot summer air, mixing in the delta with the Pacific's marine influence, creates one of the most predictably windy places in the state.

The delta is a haven for humans who like to surf the wind or catch nonnative fish such as striped bass. And while it is not a hospitable place for native fish such as the delta smelt that are protected under the ESA, there is a place in this stretch of the delta, Decker Island, where Burau can catch smelt (when conditions are just right) by the dozen.

Monitoring by the fishery agencies and researchers routinely results in the capture of more endangered smelt on an annual basis than are caught and counted (literally around the clock, every day) at the pumping facilities of the State Water Project and Central Valley Project. Burau's experience distills California's regulatory struggle with smelt. Both water agencies and smelt researchers searching for longterm solutions face severe restrictions on how many fish their very different activities can take. The restrictions have tightened over the years as monitoring has detected fewer and fewer smelt. During the biggest storm in 2012, as one example, pumping restrictions linked to smelt slowed the water projects to a near halt when available flows were at their highest. As a result, insufficient water to run the city of Los Angeles for a year was captured.

The delta environment, and the primary water systems that fuel the state economy, are locked at the regulatory hip in a decline that has yet to end.

Thanks to Burau's research, however, we now know that delta smelt and human snowbirds have something in common. They predictably travel with the weather. A favorite smelt summer hangout is just to the west of the delta in Suisun Bay. There the shallow waters of Grizzly and Honker Bays, stirred by the summer wind, create turbid water where smelt can hide from prey. Burau's research demonstrated that when the first big storm of the winter brings turbidity from the rivers to the entire delta, the smelt begin to make their move. Understanding where they go and why requires research, a netting of smelt that inevitably kills them.

To study the movement of smelt a few years ago, Burau says, wildlife agencies granted him a permit to take 600 adults over a few winter weeks. He started his monitoring at Decker Island. On the far north side of the delta's water superhighway, Decker Island has some of the most consistently turbid waters in the delta owing to the summertime sea breeze, something that these two-inch fish are smart enough to follow.

Burau and his research team monitor for smelt by towing special nets behind their aluminum-hulled boats. To start the research at Decker, "we did a five-minute tow," Burau recounts. "We caught 100 [smelt]." The State Water Project, on the other hand, entrained at the Skinner Fish Facility only ninety-two adult smelt that year. In 2016, only twenty smelt had been taken through April. The pumping projects and the pull of water in artificial directions claim other fish along the way, but undoubtedly the direct take at the pumping facilities is lower than the take from monitoring.

Burau had a four-letter-word reaction to finding so many smelt in such a small amount of time. He and fellow researchers had planned to make dozens of tows and take numerous samples over weeks. They had taken one-sixth of their allowed take of smelt in five minutes. So they moved their boats a tad upstream of Decker to tow again. "We started catching them like crazy," he recalls. "We had to bail. I was burning up my entire Endangered Species Act allocation just figuring out where they were."

Fish nets don't discriminate. They will catch anything in the water, including endangered winter- and spring-run salmon. Long before Burau's research to understand smelt migration was designed to be over, he was out of business. His nets had caught six yolk-sac salmon smolts. They were presumed to be endangered winter- and spring-run salmon. Science's ability to understand and reduce the conflict between water pumping for California and migrating patterns for smelt would have to wait for another year. "We got shut down because we caught too many winter-run-identified salmon," Burau says.

Water-pumping restrictions can be further tightened if the take of delta smelt reaches a certain level. Yet the efforts year after year to limit the mortality of smelt at the pumps has never reversed the dwindling population of the species. Within three years, the official population index had fallen another 38 percent. Monitoring in the winter of 2015 and the spring of 2016 suggested that the decline in population has continued. A spring trawl conducted by the California Department of Fish and Wildlife in April 2014 found more than 300 smelt. In April 2016 only thirty-three were found.

Finally, it is possible that the tunnels could help delta smelt in two really important ways. First, scientists believe adult delta smelt enter the central and south delta following the first big storms of the winter, where they are "salvaged" in the pumps. Salvage of adults could be almost completely avoided by shutting down the southern delta pumps and taking water into the tunnel intakes in the northern delta, where delta smelt rarely go. Second, reducing the use of the southern delta pumps during this period would cause fewer smelt to spawn in the central and southern regions of the delta. Smelt in their miniscule larval stage can be drawn into the pumps if adults spawn in this region. Shifting the spawning activity away from the southern pumps reduces another stress on the population.

The delta smelt was the first fish living out part or all of its year in the estuary to be listed under the ESA. Following its listing in 1993, it was joined the next year by the winter-run Chinook salmon. In 1999 the spring-run of salmon joined the list. In 2006 the listing was expanded to include green sturgeon and in 2010 the longfin smelt.

The biggest killer of salmon in the delta, however, is none other than predatory fish species. Predation is normally a natural part of the food chain economy in a functioning estuary. But nothing is normal about this highly altered delta. The deck has been stacked against the prey (small fish like salmon smolts and the delta smelt) and in favor of the predator, the striped bass, an introduced species, and others.

"Although smelt now are a big deal, salmon could be a bigger problem in the future," Burau says. His studies have found that young salmon that begin their lives in the San Joaquin River (the southern tributary to the delta) rarely make it to the ocean. Water project operators every year place tons of rock in the river to form barriers that prevent fish on the San Joaquin from taking a "wrong turn" at a place called Old River, which could lead them directly to the state and federal pumps, but no route from the upper San Joaquin into the delta is safe. "That barrier, it doesn't improve survival of the population," Burau says. "If you keep salmon smolts on the San Joaquin using a barrier at Old River, it just determines where they are going to die, not whether they are going to die." There is no place for a young salmon to hide, no historical habitat. The biomass of non-native plants and species such as striped bass accounts for an estimated 95 percent of all living things in the delta. The ESA, through pumping restrictions, is attempting to protect the 5 percent of delta species that belong there. It's hard to regulate a striped bass.

"They are supergood at what they do," Burau notes, in a tone of both respect and resignation. "We need to come up with habitat that is better for the prey."

Burau is not alone. A survey of delta scientists by the Public Policy Institute of California in 2012 identified habitat restoration as their highest priority.[5] But science also suggests that nature may reconfigure the delta before governments can successfully restore some of it.

Jeffrey Mount is known as the "Doctor Doom of the Delta," and for good reason. Now a retired geology professor from the University of California, Davis, he has spent an entire career studying California's rivers and both the physical and political forces that shape them. In 2005 he and Robert Twiss, a professor of environmental planning at the University of California at Berkeley, turned their focus to the delta. Rarely does a single study change the course of public policy on a topic as big as California water. This one did because of its ominous findings.

Mount and Twiss decided to study the scenario of the "big one," a northern California earthquake whose shock waves would reach the delta. At first blush, an estuary would seem to be a landscape far less susceptible to damage than an urban setting with bridges and masonry. But California set in motion some profound geological changes in the delta when the state essentially decided to occupy the estuary by carving it into islands with 1,100 miles of levees.

Core samples referred to by Mount and Twiss in their 2005 study suggested that the delta has been a delta for about 6,000 years.[6] And until the construction of the levees, the delta was in a state of natural balance in terms of elevation and sediment. The thickets of tules created a rich peat soil and other sediments. Tides and the occasional floods would redistribute those and other sediments arriving in flows from upstream. The delta was an ever-changing marshland, but it was an ecologically functioning one.

Then, in the name of reclamation, the delta was carved into pieces by the levees. And the lands were drained by elaborate ditch systems built by farmers to expose and prepare these peat soils for cultivation. Steadily, a very different natural process began to happen.

"Draining tule marsh soils initiated a sustained period of land subsidence that continues today," Mount and Twiss wrote. "Today, the Delta is a mosaic of levee-encased subsided islands with elevations locally reaching more than 8 meters below sea level."[7]

Peat soils oxidize as they are exposed to the atmosphere. Delta farmers once literally burned some soils, accelerating the process. Grading the soils can further accelerate the loss of these rich soils. As the islands have lost their soils, they have declined in elevation and remain dry and farmable only because of the levees, under constant test by the delta waters passing by.

Mount and Twist estimate that decomposing tules and other natural processes created about 5 billion cubic meters of peat soils in the 6,000-year history of the delta. In just 100 years of human occupation of the delta, half these soils are gone. That has created 2.5 million cubic meters of island subsidence in the delta. The subsidence in the delta is the equivalent in total depth to about 8,000 empty Rose Bowls.

Researchers like Mount and Twiss refer to these holes as "accommo-dation space." It is a void begging to be filled by one natural process or another. In the delta, nature offers two disaster scenarios, a flood or an earthquake.

"When levee breaches occur on deeply-subsided islands, rapid filling draws brackish water into the Delta, temporarily degrading water quality over a large region," Mount and Twiss wrote. "Known colloquially as 'The Big Gulp,' the water quality impact of island filling is principally a func-tion of the magnitude and location of anthropogenic accommodation space."[8]

As a consequence, widespread levee failures could put the nation's two largest water projects out of commission because saltwater from San Fran-cisco Bay would rush eastward to fill the void in the delta. "If regional is-land flooding results in numerous levee breaches, it is unlikely that levee integrity can be restored for many years, with protracted disruption of water supply and loss of farm income," they wrote. It is no accident that this scenario is known as the Big Gulp.

How likely is this scenario?

"There are at least five major faults within the vicinity of the delta capable of generating peak ground acceleration values that would likely lead to levee failures," they wrote. They evaluated the impact of a 100-year flood and a 100-year earthquake and the probability of either disaster happening in a fifty-year time interval. Their conclusion: "a roughly two in three chance." "This discussion is meant to highlight the fact that punctu-ated landscape change in the Delta is not a remote, hypothetical possibil-ity, but is highly likely during the simulated period of 50 years." They concluded, "In our view, there is no comprehensive scientific effort to address this issue and to provide the necessary information to inform policymakers."[9]

A Durable Vision: The Two-Track Solution

The study was released in March 2005. Within the next eighteen months, then governor Arnold Schwarzenegger had commissioned a blue ribbon task force to develop a "durable vision" for the delta, citing the seismic risk. The Metropolitan Water District of Southern California established criteria for a long-term delta solution that addressed seismic risk. And state and federal agencies, along with public water agencies, signed a planning agreement to officially launch the BDCP process in an effort to find a comprehensive solution for both the water system and the delta ecosystem.

Speaking later to *Popular Mechanics*, Mount acknowledged the engineering solution to the water-supply problem identified by his research. For capturing and moving water supplies, "You can now engineer a structure that is earthquake resistant, resistant to the effects of sea level rise, subsidence and changing inflows."[10] This led to the concept of building tunnels beneath the delta, which was essentially an updated version of the peripheral canals Jerry Brown had proposed in 1980. And this time the plan included habitat restoration to address the intractable conundrums Jon Burau focuses on. "The combination of the tunnels and habitat could be a great thing," Burau says. "They can work synergistically together. They have to be done together."

But could a BDCP actually happen as in a single grand plan for a delta water system and ecosystem anchored for fifty years through state and federal permits? The short answer appears to be no.

By the end of the summer of 2014, state and federal agencies had received 12,000 public comments on the grand proposal of the BDCP. Among them was a particularly harsh letter from the U.S. Environmental Protection Agency. There remained considerable resolve by both state and federal administrations and public water agencies to move forward. The question was how.

The idea of a grand plan, the BDCP, began to fade. The notion of incremental progress on all the same fronts began to take hold.

Permits with a fifty-year horizon appeared beyond the reach of the agencies that would have to issue them. The delta is a landscape of uncertainty on every conceivable front. There are rising temperatures and evolving conditions resulting from climate change. And there is simply not enough habitat that is already restored to know, with regulatory certainty, the precise benefits of restoration.

Mark Cowin, director of the California Department of Water Resources, explained the strategy shift to the media in the summer of 2015 this way:

> We began to have doubt as to whether or not a 50-year habitat conservation plan is realistic, primarily given the great uncertainty about future ecological conditions in the Delta under climate change. Complicating the matter is the fact that the state and federal regulators that would provide the permits face a lack of scientific data about how the Delta's estuary might respond to large scale habitat restoration. We know that something on the order of 90 percent of native habitat in the Delta has been lost due to reclamation of those Delta islands, but what we don't know in detail is just what kinds of habitat restoration, where the habitat restoration should be located, and how all of that together will help the fish populations recover and just how successful that recovery would be.[11]

By April 2015 the BDCP was gone as the official government moniker of the delta strategy. The BDCP process remained the vessel to ferry the environmental documentation to the finish line, but in essence, the solution was split into two separate efforts, to proceed on independent yet parallel tracks. The water system solution—the intakes and the tunnels—stayed essentially the same and has become known as California Water-Fix. The restoration solution became known as California EcoRestore. Its time horizon was shrunk to about five years, with an ambitious target of 30,000 acres of enhanced wetlands and floodplains.

The challenges are many. Wildlife agencies must have the courage to issue new permits despite worsening ecological conditions and scientific uncertainty. Water agencies must have the courage to make an estimated $15 billion in delta investments without a rock-solid certainty of outcomes. Political leaders must lead, putting statewide needs over local resistance. In short, California and the federal government must do something when it comes to the delta that they have never managed to do before.

"The very fabric of a modern California is at stake," Governor Brown told reporters in April 2015 on the shift of delta strategy to California

WaterFix. "I'm doing what I believe is absolutely necessary to proceed forward."[12]

He may be understating the importance.

California is not the only western state with skin in this delta drama. Thanks to massive diversion projects, the Colorado River basin is connected to many of the west's major urban centers: Denver, Salt Lake City, Albuquerque, Phoenix, Las Vegas, and more than 20 million residents of Southern California. The Colorado River feeds the fifth-largest economy in the world. It is an integral part of a massive, West-wide artificial watershed that extends from Cheyenne to San Francisco, touching everything in between before it spills into northwestern Mexico and becomes a primary water supply to that part of our neighbor to the south. A failure in the delta would be felt far beyond Los Angeles, San Diego, Silicon Valley, and Bakersfield.

As the Colorado River community struggles to adapt to increased droughts, it cannot also carry the burden of a "broken" delta. The connection between these two systems became readily apparent in 2014, when the Metropolitan Water District of Southern California, having been cut off from virtually all water resources from the delta, had to rely on water it had acquired and stored in Lake Mead on the Colorado River. But 2014 was also one of the worst drought years ever experienced within the Colorado River system. So a reservoir that was already declining significantly because of bad hydrology was forced into further decline because Southern California had no other option available to it. For this larger multi-state, multinational community, finding a path forward in the California delta is critical to a larger regional solution to the water-supply problem.

Cooperation under a Regime of Change

The effects of climate change are descending on the western United States at a pace few had thought possible. Everything regarded as improbable or as belonging to the distant future has become a present reality. All over the world, similar scenarios are unfolding. California has an opportunity to stand as a beacon for new age water solutions in an ever-changing landscape. But it can do so only if narrow self-interests can be integrated into a larger regional solution. It is the last opportunity to find a new fulcrum, a new balance point for urban, agricultural, and environmental requirements.

Human hands created the conundrum. Only human cooperation and mutuality can carve a pathway toward a more sustainable coexistence. If they fail, the delta will once again become a tragic and lost opportunity, with consequences that are difficult to imagine.

Notes

1. Aaron Sankin, "Jerry Brown: 'I Just Want To Get Sh*t Done,'" *Huffington Post*, September 4, 2012.

2. San Francisco Estuary Institute-Aquatic Science Center, "A Delta Transformed: Ecological Functions, Spatial Metrics, and Landscape Change in the Sacramento-San Joaquin Delta," Publication #729 (Richmond, CA: San Francisco Estuary Institute-Aquatic Science Center, 2014), p. 20 (www.sfei.org/projects/delta-landscapes-project).

3. Ibid.

4. Ibid.

5. Ellen Hanak, Caitrin Phillips, Jay Lund, John Durand, Jeffrey Mount, and Peter Moyle, "Scientist and Stakeholder Views on the Delta Ecosystem," Public Policy Institute of California Study, April 2013 (http://www.ppic.org/content/pubs/report/R_413EHR.pdf).

6. Jeffrey Mount and Robert Twist, "Subsidence, Sea Level Rise, and Seismicity in the Sacramento–San Joaquin Delta," *San Francisco Estuary and Watershed Science* 3, no. 1 (2005), Article 5 (http://repositories.cdlib.org/jmie/sfews/vol3/iss1/art5).

7. Ibid., p. 7.

8. Ibid.

9. Ibid.

10. "5 Questions for Geologist Jeff Mount on Californias Crumbling Delta Levees," *Popular Mechanics*, October 30, 2009 (http://www.popularmechanics.com/science/a2782/4258291/).

11. "Director Mark Cowin on the Revised Environmental Documents for California Water Fix, Formerly Known as BDCP, More Commonly Known as the Delta Tunnels," Maven's Notebook, July 13, 2015 (https://mavensnotebook.com/2015/07/13/media-call-director-mark-cowin-on-the-revised-environmental-documents-for-california-water-fix/).

12. David Siders and Phillip Reese, "Jerry Brown's Revised Water Tunnels Plan Adds Political Problems," *Sacramento Bee*, April 30, 2015.

Chapter 3

The San Diego Strategy
A Sea Change in Western Water

MAUREEN A. STAPLETON

As drought squeezed California and the West harder and harder from 2012 to 2016, the largest seawater desalination project in the Americas rapidly took shape along the glittering shoreline of northern San Diego County—a futuristic fast-forward in the region's unrelenting quest for reliable water supplies that goes back to the first Franciscan mission built in California nearly 250 years earlier.[1] With few rain delays, construction on the $1 billion Carlsbad Desalination Project moved well ahead of schedule to produce 50 million gallons a day (mgd) of drought-proof water for the San Diego County Water Authority and its twenty-four member agencies. Owing to the extraordinarily hot and dry conditions, the plant could not have been timelier or more welcome when commercial production started in December 2015. State drought regulators quickly lowered mandatory water-saving targets for the San Diego region because of the new drought-resilient supply from the facility, providing an immediate benefit for the region's timely investment.

Despite appearances, the desalination project wasn't conceived as a response to the drought. Indeed, its origins go back to the early 1990s, when California had just emerged from another epic dry spell and the Water

Authority had initiated a search for a diversified water-supply portfolio that would protect it from overreliance on a single provider of imported water. Until that point, the Water Authority was mainly focused on purveying imported water through large-diameter pipes to member agencies, which in turn served San Diego County's growing economy and population. Nearly the entire supply, 95 percent, came from the Los Angeles–based Metropolitan Water District (MWD) of Southern California. As that MWD's supplies shriveled in 1991, water deliveries to San Diego County were cut by 31 percent, damaging the region's then $65 billion economy and the quality of life for 2.5 million residents. What wasn't evident then but seems clear in hindsight is that drought conditions from 1987 to 1992 were part of a trend toward a hotter and drier climate. The changing climate in turn has forced water agencies across the West to rethink the generations-old model for meeting water demands largely through reliance on the annual snowpack and surface water storage.

From the crucible of the 1987–92 drought there emerged a multidecade strategy by the San Diego County Water Authority to improve the reliability of the region's water supply and a determination to fundamentally change course by developing numerous climate-resilient resources through conservation and the use of water transfers, recycled water, groundwater, and desalinated seawater. Almost twenty-five years later, it is one of the most aggressive and well-executed water plans of its kind in the United States and a model for how community resolve can turn desperation into diversification. Business leaders understood the importance of the strategy and embraced it, providing pivotal endorsements for costly investments, and the Water Authority itself emerged as an important state and regional force in water-supply development and conservation efforts, no longer content to simply take whatever came down the pipe.

The Water Authority has cut reliance on its largest supplier by more than half, and that figure will continue to shrink at least through 2020 as new water sources come online. To make seawater desalination work, the Water Authority not only had to pioneer a new financing model for major municipal water infrastructure projects, it also had to reengineer a delivery system designed decades ago to move water from distant mountains to population centers on the coast. It is a sea change that offers hope for other arid coastal regions seeking new sources of reliable supply in an era of continued population growth and a fast-changing climate.

Drought Prompts Diversification

When the drought conditions that struck California in the late 1980s persisted into the next decade, the water-supply crisis gained momentum with remarkable speed. In less than three months, between late 1990 and early 1991, MWD sped through its drought response stages, moving from stage 1 to stage 5 and imposing a combined 31 percent cutback for municipal and agricultural customers in the San Diego region. On March 4, 1991, it added another stage, stage 6, which would amount to a 50 percent water-supply reduction for the San Diego County Water Authority's service area and a devastating blow to the county's $1 billion farm industry, its nascent life sciences industry, and the rest of the economy. Figure 3-1 shows the progression of the MWD's drought stages during the pivotal six-month stretch.

As might be expected, the region's business and civic leaders demanded to know why the San Diego County Water Authority had virtually all of its water-supply eggs in one basket. During public meetings they lined up for city blocks to tell the board of directors to develop a better plan. Respite for the region came in spades from nature during what is still called "Miracle March." Some seven inches of rain fell that month and took the edge off drought conditions.[2] Nevertheless, a 31 percent reduction in imported water-supply deliveries remained in effect and would last for more than a year.

The Water Authority took immediate actions in response to the supply allocations, first boosting emergency deliveries through dry-year transfers from the state's drought water bank and pushing for increased conservation by residents, businesses, and farmers. Conservation became the foundation block of the emerging diversification plan because it was a tactic that could be deployed quickly. In 1990 the San Diego region used 235 gallons per capita per day (gpcd). Two years later that amount was

FIGURE 3-1

Shortage Allocations of the Metropolitan Water District of Southern California during Drought Progression, 1990–92: Stages of the Interim Interruptible Conservation Program (Shortage Allocation Plan)

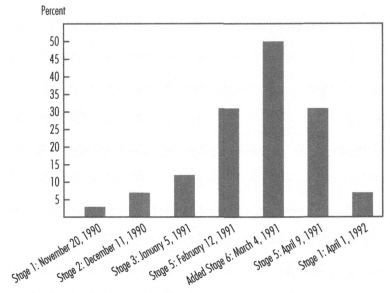

Source: San Diego County Water Authority.

down to 175 gpcd as the region responded to emergency conditions and outreach efforts around the agency's "Don't Be a Water Hog" campaign.

As necessary as the immediate reduction in water use was, the Water Authority realized more measures would be needed to ensure conservation practices would endure for the long term. Along with more than 100 other water agencies and environmental groups, it helped form the California Urban Water Conservation Council, with the goal of maximizing urban water conservation through technological innovation, effective policies, and public education. Another prong of the effort to reduce water use involved offering free home water-use evaluations along with financial incentives to boost device-based conservation. The regional effort was huge because of the convergence of the drought with compliance efforts to reduce wastewater discharges into the ocean. Across San Diego County, more than 1.2 million low-flow toilets, water-saving showerheads, and high-efficiency clothes washers have been installed since 1990 through the Water Authority's conservation programs.

The Water Authority also pushed for fundamental and far-reaching changes at the state level, and its legislative efforts resulted in new statewide standards for the 1.6 gallon per flush toilet, which quickly became the de facto national standard as manufacturers catered to the California market. It was formalized in federal policy as part of the Energy Policy Act of 1992. "National water savings attributable to 1.6-gpf toilets and other low-volume fixtures required by [Energy Policy] Act 1992 are and continue to be unprecedented. An estimated 7 bgd [billion gallons per day] of water—enough to supply seven cities the size of New York City—are being saved in the United States," according to a 2014 article.[3] Over the years, additional state legislation sponsored or supported by the Water Authority led to water-efficient landscape ordinances, statewide water metering, and the establishment of a 20 percent per capita water-use reduction goal statewide by 2020, effectively hard-wiring conservation practices and targets for California.[4] The results were significant: conservation efforts reduced the San Diego region's water demands by 73,000 acre-feet in 2014. Since 1990, San Diego County has added more than 800,000 residents and expanded its gross domestic product more than 90 percent (adjusted for inflation) while using 33 percent less potable water.

Regional success in reducing potable water use is partly attributable to water agencies in San Diego County developing wastewater recycling projects in the 1990s to take the pressure off drinking water supplies. Large-scale recycling and groundwater recharge efforts already were under way to the north in Los Angeles and Orange counties but had not become widespread in San Diego County, with its relatively small aquifers.[5] Drought-fueled efforts in the 1990s, coupled with a regional push to reduce wastewater discharges into the ocean, led to the expansion of a purple pipe distribution system to service heavy nonpotable water users such as golf courses, farms, and business parks with landscaping. Those projects were completed by the Water Authority's member agencies where conditions were favorable, for example, where wastewater plants were near plant nurseries. The Water Authority helped secure grant money for local projects and to fund a study of advanced recycling methods for potable water reuse by the National Research Council that validated the technology. That was followed in 2013 by successful Water Authority–sponsored state legislation to expedite a rigorous and transparent scientific assessment of potable reuse as a potential water source.[6] Water recycling in the San Diego region generated 26,000 acre-feet in 2015—five times more than all the surface water produced in the county that year—and the volume of water that gets used more than once is expected to expand significantly as larger-scale projects involving potable reuse move from concept to reality.

The other major element of the Water Authority's diversification plan that took shape in the 1990s was water transfers from the Colorado River, some 150 miles east of the metropolitan San Diego area. Starting in 1995, the Water Authority and the Imperial Irrigation District explored ways to conserve and transfer water from inland agriculture to the urbanized coastal plain. The discussions were prompted not just by the Water Authority's desire to diversify its supply but also by the Imperial Irrigation District's desire to fund improved water-use efficiency measures and protect its senior water rights. A Water Conservation and Transfer Agreement signed by the parties in 1998 envisioned up to 300,000 acre-feet a year (the volume eventually was set at 200,000 acre-feet), and a related agreement provided 80,000 acre-feet a year of conserved water from canal-lining projects in the Imperial Valley to San Diego County. Those documents are part of the Colorado River Quantification Settlement Agreement, which was signed in October 2003 at Hoover Dam by federal, state, and local officials. The agreement comprises several significant elements, including a plan to limit California to its basic annual apportionment of Colorado River water (4.4 million acre-feet). The Water Authority-Imperial Irrigation District agreement remains the largest farm-to-city water transfer in U.S. history, at an estimated 7 million acre-feet over forty-five years. Today the Colorado River transfers meet more than a third of the San Diego region's water demand, and the percentage is expected to grow to 46 percent by 2020. Figure 3-2 shows how the Quantification Settlement Agreement water transfers ramp up between 2003 and 2021.

The major new piece of the region's supply portfolio is seawater desalination, a resource that has been discussed, debated, and developed in San

FIGURE 3-2

Imperial Irrigation District (IID) and Canal-Lining Deliveries, 2003–21

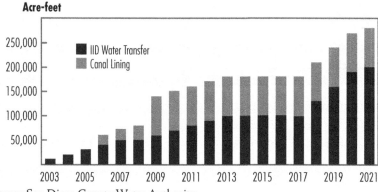

Source: San Diego County Water Authority.

Diego for decades but is only now a reality for the Water Authority and its member agencies.

Desalination Concept Emerges

General Atomics has had a significant influence on San Diego County and the nation during the last half of the twentieth century with major contributions to nuclear energy, space exploration, and national defense. Less well known is the San Diego–based company's research in the 1960s to engineer and commercialize a water purification system called reverse osmosis for the U.S. Department of the Interior's Office of Saline Water. It was an attempt to reach the holy grail of water production envisioned in the 1790s by Samuel Taylor Coleridge in the "Rime of the Ancient Mariner": tapping the inexhaustible ocean for drinking water. Indeed, President John F. Kennedy in 1962 coined what has become something of a mantra in the water industry when he said that the low-cost production of freshwater from saltwater would be such a great service to humanity that it would "dwarf any other scientific accomplishment."[7] Kennedy was also quoted in 1961 as saying, "No water resources program is of greater long-range importance than our efforts to convert water from the world's greatest and cheapest natural resources—our oceans—into water fit for our homes and industry. Such a break-through would end bitter struggles between neighbors, states and nations."[8]

At the same time that General Atomics was pioneering and patenting its spiral-wound RO technology, the Office of Saline Water was selected by the Navy to provide its base at Guantánamo Bay, Cuba, with freshwater after the water-supply line from Cuba was severed. In 1964 the Office of Saline Water relocated a steam-powered "flash evaporator" desalination plant from a demonstration site at Point Loma, on the San Diego County coastline, to Cuba. The evaporator technology was eventually replaced with RO membranes, and the base's desalination plant has been churning out freshwater ever since.[9]

Despite its local origins, desalination developed much more slowly for civilian purposes in San Diego County, even after interest was revived by the drought of 1987–92. At the time, RO technology was still improving and maturing; the energy-intensive process involved forcing seawater through tightly wrapped membranes under very high pressure so that water molecules could pass through but salts could not. Despite its intriguing possibilities, the technology remained far too expensive compared to the region's historical supplies from the Rocky Mountains and the Sierra Nevada. In addition, it would take much longer to design, permit, and build a desalination plant than to increase conservation or even to build nonpotable water recycling projects in response to drought conditions. But as the Water Authority gazed into the future in a 1993 study, the vision for desalination started to grow. Two potential sites emerged from that report, both next to power plants on the San Diego County coastline. The power plants at Carlsbad and Chula Vista offered not only proximate sources of energy but also water-supply infrastructure, since both relied on existing ocean intakes for cooling. The Encina Power Station site in Carlsbad was the more viable location because of concerns about expensive repowering upgrades needed at the Chula Vista facility and questions about the plant's long-term viability. Both sites raised significant logistical questions about how this new water source at sea level would be integrated into the regional delivery system, which relied on parallel aqueducts running through inland valleys ten to fifteen miles to the east. Despite such complications, desalination was officially recognized as a potential water supply for the San Diego region in the Water Authority's 1995 Urban Water Management Plan. For the rest of the decade, however, the agency's primary focus would be on securing independent Colorado River water transfers though complex multiparty negotiations.

Sensing the potential for a booming desalination business in heavily populated Southern California, privately owned Poseidon Resources (now Boston-based Poseidon Water) arrived in San Diego County in 1998 with a model from the electricity industry in which private developers generate a resource and sell it at a profit to "off-takers" such as utility companies. The concept was foreign to public water agencies in the West, where the vast majority of water supplies over the preceding century had been delivered by federal, state, and local public water providers at cost. Despite the potential disadvantages, desalination appeared to be ready to make a run in the San Diego region by 2001: the Water Authority's board had approved a Desalination Action Plan, Poseidon had locked up a long-term lease at the Encina Power Station site, and the city of Carlsbad had invited the Water Authority to take the lead in negotiations with Poseidon.

The parties entered a term sheet in 2002, setting the stage for a desalination deal as Poseidon was developing a 25 mgd desalination plant in Tampa Bay, a plant that was supposed to open in 2003 but was haunted for years by cost overruns and operational delays. Entanglements in Tampa Bay raised concerns about the ultimate cost of water in Carlsbad and Poseidon's ability to deliver water from a project in San Diego County. Other complications emerged regarding the transparency of the contracts and other documents, along with the long-term return on investment for water ratepayers, that couldn't be resolved to the Water Authority's satisfaction. There were, essentially, deep cultural differences between the public and the private sectors that could not be bridged. The Water Authority wanted transparency; Poseidon claimed trade secrets and confidentiality. The Water Authority sought the highest return for ratepayers; Poseidon served investors. Negotiations were suspended in 2003, and the parties went their separate ways without giving up the hope that desalination would eventually emerge as a part of the region's water supply. In 2005 the Water Authority's board again identified desalination as an important new source of water supply (as it had done in 2000 and 2003 and would do again in 2010) and set a goal to develop capacity for 56,000 acre-feet a year of seawater desalination by 2020.

With the Chula Vista desalination option off the table because of the power plant's expected closure (and the attendant loss of the seawater used for cooling), the Water Authority looked to the very northern edge of the county and studied the concept of a desalination plant colocated with the San Onofre Nuclear Generating Station, which also had large seawater intakes for cooling.

The Water Authority started discussions with the Marine Corps' Camp Pendleton, a member agency with a 125,000-acre installation, seventeen miles of coastline, and tens of thousands of thirsty troops. Those talks led to a feasibility study in 2009, which was followed by technical studies the next year to evaluate building and operating a desalination plant that could generate 50–150 mgd on the base. (Prior studies had shown that 50 mgd was the sweet spot for economies of scale.) At the same time, the Water Authority continued assessing the potential for a plant on Carlsbad's shoreline, going so far as to conduct its own environmental reviews at the Encina site despite potential limitations such as a compact footprint for work bounded by the power plant, a rail line, and a lagoon. And the agency sought public input about desalination through its regular public opinion surveys in 2006, 2008, and 2010. Each time, whether the region was in drought or not, more than 80 percent of respondents said seawater desalination was an important piece of local water-supply reliability efforts. Public enthusiasm was coupled with caution inside the Water Authority because serious questions remained about who would operate Poseidon's plant, how the plant would connect to existing infrastructure, and the ultimate cost of water obtained this way.

Meanwhile, Poseidon also performed extensive environmental analyses at its leasehold site in Carlsbad as it moved doggedly through a complex regulatory process. Over a six-year stretch, the company obtained twelve local discretionary approvals and seven discretionary approvals from state agencies, including the California Coastal Commission, not to mention the fourteen environmental lawsuits in which Poseidon prevailed. The company also solicited and received the unanimous backing of the region's state and federal legislative delegations. Despite those victories, Poseidon couldn't finalize a project financing structure with nine local water agencies (including the host city of Carlsbad) that had indicated interest in buying water directly from the proposed plant. The impediment boiled down to cost: while desalinated seawater priced out about the same as other new sources of drought-proof supplies, such as water reuse or brackish groundwater desalination, it was roughly twice as expensive as conventional imported water supplies. Local agencies and Poseidon negotiated a pricing concept in which the company would absorb some of the costs in the early years so that the price of desalinated water was no more than the cost of imported water from the Water Authority. When that proved infeasible, the retail member agencies balked at the fiscal impact on their relatively small customer bases and in 2010 asked the Water Authority to reconsider whether an agreement between the Water Authority and Poseidon made sense for the region as a whole by spreading costs across some 3 million residents and an economy nearing $200 billion in gross domestic product.

By that time, California was mired in another drought, and interest in additional water sources was swelling. Not only that, but the science of climate change was making it increasingly clear that weather extremes would grow and the hydrograph that western water agencies had relied on for decades was shifting. Scientists at the Scripps Institution of Oceanography at the University of California, San Diego, were among the world leaders in studying the issue. In the 2004 paper "Recent Projections of 21st-Century Climate Change and Watershed Responses in the Sierra Nevada," Michael Dettinger and others warned of "important changes in precipitation, extreme weather, and other climatic conditions, all of which may be expected to affect Sierra Nevada rivers, watersheds, landscapes, and ecosystems" that historically have provided San Diego County and Southern California with significant amounts of water.

Even the modest climate changes projected by the [Parallel Climate Model] (with a conservative value for warming and small precipitation changes) would probably be enough to change the rivers, landscape, and ecology of the Sierra Nevada, yielding: (1) substantial changes in extreme temperature episodes, for example, fewer frosts and more heat waves; (2) substantial reductions in spring snowpack (unless large increases in precipitation are experienced), earlier snowmelt, and more runoff in winter with less in spring and summer; (3) more winter flooding; and (4) drier summer soils (and vegetation) with more opportunities for wildfire.[10]

A 2008 study, also by Scripps researchers, gave 50 percent odds that "live storage" in Lake Mead, a key source of water for San Diego County and tens of millions of people across the Southwest, would be gone by 2021 if the climate changed as expected and future water use wasn't curtailed. "Time is short," the report concluded. "The alternative to reasoned solutions to the coming water crisis is a major societal and economic disruption in the desert southwest."[11] The San Diego County Water Authority had already begun building capacity to handle drought and other emergencies (such as an earthquake severing critical imported water-supply lines) through its $1.5 billion Emergency & Carryover Storage Project. That initiative included Olivenhain Dam, the region's first major new dam in fifty years, and the nation's tallest dam-raising project at San Vicente Dam. That monumental effort created 152,000 acre-feet of additional storage for dry years and emergencies, the largest single increase in reservoir storage in county history, as part of the Water Authority's larger Emergency & Carryover Storage Project.

While those upgrades were necessary for improving the storage and movement of water around San Diego County, they didn't generate any new supplies. To complicate matters, the Water Authority had neither a lock on a site for its own seawater desalination project nor experience in designing and constructing such a project. For that, the Water Authority's board again decided to seek terms with Poseidon at the urging of several member agencies (which have seats on the Water Authority's board of directors) in 2010. Significantly, Poseidon had secured all necessary permits for the plant by that time, and it had selected IDE Technologies, a worldwide leader in the design, construction, and operation of desalination plants, as the desalination process contractor for the Carlsbad plant through a competitive proposal process. IDE's proven track record with major RO facilities in Israel and elsewhere gave the Water Authority far greater confidence in the prospects for a viable desalination project than it had had eight years earlier.

Landmark Agreement Takes Novel Approach

In July 2010 the Water Authority's board approved key terms for purchasing water from Poseidon, setting the stage for detailed negotiations. Besides the ultimate price of water, the biggest issue for the Water Authority was the ability to shift risk for potential construction cost overruns, slowdowns, and nonperformance of the plant to the private sector. With the Tampa Bay problems as the backdrop, the Water Authority was determined to put the responsibility and liability for design, permitting, financing, construction, and operation of the project onto Poseidon and its investors. The Water Authority also made it clear from the start of the formal talks that it would require a thirty-year term with buyout options to effectively guarantee continual upkeep of the plant and public ownership after three decades for $1. And the Water Authority demanded oversight capacity during both project construction and operations. It would not be good enough, the Water Authority insisted, to get the water; it needed to know how the water was produced and that the highest standards were met throughout the process. Even with those foundational elements in place, the Water Authority required that before starting direct negotiations on a water purchase agreement, Poseidon terminate all confidentiality agreements with the Water Authority and submit a binding commitment for construction from an equity investor. Also, all nine Water Authority member agencies that previously had water purchase memoranda of understanding with Poseidon would have to cancel those agreements in writing.

It took more than a year for those conditions to be met, plus another year of detailed—and often tense—negotiations to draft what would end up being a 222-page Water Purchase Agreement (with eighteen technical appendices) for public and board review.[12] During negotiations, Water Authority staff worked with Poseidon to complete planning and technical

FIGURE 3-3

Desalination Project Components

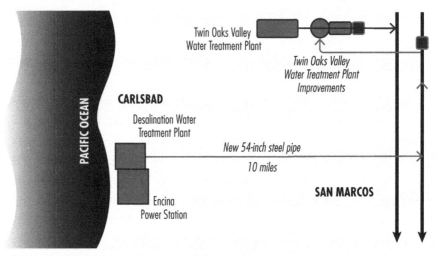

Source: San Diego County Water Authority.

studies related to the capital improvements that would be necessary to deliver and integrate water from the desalination plant into the Water Authority's regional water storage and pipeline system. Another critical element of the eventually successful enterprise was Poseidon's new leadership and investors, who brought a fresh view and willingness to compromise.

The result of the marathon talks, made public in September 2012, was a three-part project: the desalination plant; a ten-mile, fifty-four-inch-diameter conveyance pipeline to connect the plant to the Water Authority's existing aqueduct; and upgrades to the Water Authority's existing infrastructure. Figure 3-3 shows a simplified schematic of the desalination project components.

According to the terms of the agreement, the Water Authority would have the option—but not the obligation—to buy the $537 million plant after the first decade of operation using a preset price formula and the opportunity to buy the plant for $1 after thirty years. As for the $159 million pipeline, the Water Authority would own and operate it once commercial water production began at the plant, but if Poseidon failed to deliver water it would have to pay the pipeline's debt service. This contractual requirement was part of the agency's effort to protect ratepayers from the risk of building a "pipeline to nowhere" if the plant proved faulty. Improvements to the Water Authority's infrastructure were pegged at $80 million, and project financing costs (such as debt service reserve accounts, interest during construction, and transaction fees) were an estimated $227 million, bringing the total project to just over $1 billion.

The pioneering Water Purchase Agreement effectively balanced private and public interests, with neither side in complete control. The core principle of the partnership proved solid: the Water Authority would only pay for water actually produced by the plant that met predefined quality and quantity specifications; Poseidon, in exchange for taking on a significant amount of the risk, would get a slightly higher price for its product water. (In contrast, the Tampa Bay plant was financed with full credit recourse to the public water agency after it purchased the troubled project.) But the risk transfer strategy had its limits; the private sector wouldn't be left shouldering all the risk. In the Carlsbad case, risk for a significant element of the operating costs, electricity, was split between the parties, with the Water Authority agreeing to assume the risk of electricity price increases over time (as it does with all of its own facilities) and Poseidon assuming the risk of electricity consumption increases. That arrangement provided Poseidon with an incentive to limit energy use, even prompting the adoption of state-of-the-art energy-recovery devices, which came with the indirect benefit of reducing carbon emissions caused by the generation of electricity for the project. Overall, energy costs account for about 27 percent of the all-in cost of water from the project. Other elements of the agreement gave the Water Authority the right to ensure the plant is operated in a safe and efficient manner, including the ability to set employment standards for key personnel, establish reporting and record-keeping requirements, and perform inspections. Poseidon also agreed to maintain the plant in conformance with industry standards, including repairing and replacing equipment according to a preset schedule. And the company agreed to pay any costs greater than $20 million (adjusted for inflation) for upgrading the power station's seawater intake to state regulatory standards when it's no longer used for power production (expected in 2017).

The Water Purchase Agreement set the price of water at $1,849 to $2,064 per acre-foot (2012 dollars), depending on the annual purchase volume. The first 48,000 acre-feet bought by the Water Authority each year cover the fixed costs of the project and the variable costs of producing water.[13] The Water Authority has the option to purchase an additional 8,000 acre-feet per year at a lower rate that reflects only the variable costs of incremental water production. The arrangement incentivizes the Water Authority to maximize plant output because any amount over 48,000 acre-feet is less expensive. The total cost of water, after accounting for the purchase price from Poseidon and the cost of the Water Authority's own facility modifications, ranges from $2,014 to $2,257 per acre-foot (2012 dollars), depending on purchase volumes.

While the impact on ratepayers varies by local water agency, typical monthly costs are about $5 per household for the first full year of desalinated water deliveries in 2016, at the low end of the Water Authority's 2012 forecast.

Annual payments to Poseidon by the Water Authority will increase by an estimated 2.5 percent per year to account for inflation and debt service. This approach also allows the cost of desalination to be phased in, ensuring that both current and future ratepayers help pay for the project. In addition, Poseidon can increase its price to accommodate changes in law or regulations that apply industry-wide to water treatment and desalination facilities. Over the thirty-year term, cumulative increases are capped at 30 percent. While the costs of desalinated seawater are relatively high at the start of production, Water Authority projections show that the price of desalinated seawater will be comparable to the price of supplies from the MWD of Southern California as soon as the mid- to late-2020s, creating a long-term supply source that likely will produce positive rate effects in little more than a decade.

The minimum annual water purchase required by the Water Authority accomplished two vital goals: it provided Poseidon with assured sales of 48,000 acre-feet (provided water from the plant met standards) that it could take to the bond market to finance the project, and it provided the

Water Authority with an amount of water that would be needed regardless of the weather. Determined not to pay for a stranded asset in Carlsbad, the Water Authority calculated numerous supply-demand scenarios that included very wet years, when water purchases would shrink. The Water Authority concluded there would be regional demand for at least 48,000 acre-feet even in years when rain curtailed consumption. At the upper end of the purchase agreement, 56,000 acre-feet reached plant capacity, given the physical limits on the intake system and membranes.[14] The result was a facility that would become a full-time core supplier for the San Diego region and reduce dependence on imported water supplies year in and year out, while proving especially valuable in dry times.

Though it's often overlooked, the public process that played out between 2010 and the signing of the Water Purchase Agreement in November 2012 was critical to the project's success. While the agreement was being developed, the Water Authority's board held twenty-six public meetings to discuss various aspects of the contract and the project as a whole. After the purchase agreement was released to the public in September 2012, the Water Authority hosted six more public meetings or workshops to discuss specifics of the agreement. In light of favorable public opinion polls and overwhelming political and civic support, the Water Authority's board approved the contract in November 2012 with a clear sense of public sentiment that tilted strongly in favor of moving forward with the desalination project despite the additional costs.

That's not to say that everyone loved the project. Environmentalists opposed it from the start and tried their best to derail it during the permitting process and through litigation. Aside from their concerns over the plant's environmental impact, none of which was deemed significant enough to stop the project, the main point of contention raised by environmentalists was that there were cheaper and better alternatives. Specifically, they pressed for more conservation and water recycling, two strategies that the Water Authority and its member agencies had invested in for years before the Carlsbad Desalination Project started production and will continue to aggressively support. In addition, Poseidon made significant environmental upgrades during the process, boosting its energy efficiency, offsetting greenhouse gas emissions, and enhancing coastal habitat.

The suite of enhancements made the Carlsbad Desalination Project one of the most environmentally friendly projects of its kind in the world. For instance, it is on track to be the first major California infrastructure project to eliminate its carbon footprint through the purchase of carbon offsets and energy-recovery devices. Poseidon is also restoring sixty-six acres of habitat along the San Diego County coastline as a marine life mitigation measure and is assuming responsibility for the continued stewardship of Agua Hedionda Lagoon adjacent to its facility.

Before construction could begin, the Water Authority and Poseidon needed to clear one more hurdle, financing on the bond market, that likely would not have been possible without the collaborative spirit that grew during contract negotiations. It was a complicated endeavor involving agreements between Poseidon, its equity investors, the San Diego County Water Authority Financing Authority, and the California Pollution Control Financing Authority, a conduit issuer of low-cost bonds for qualified projects. Tax-exempt bonds were issued by the California financing authority, with proceeds from the plant bonds loaned to Poseidon for construction of the desalination plant and proceeds of the pipeline bonds loaned to the Water Authority's financing agency for the conveyance project. The financing structure created an interest rate saving on the bonds, plus a saving on taxes and franchise fees that a private party would have to pay. Overall, the savings that emerged from the Water Authority's ownership of the conveyance pipeline were estimated at $27 million on a present value basis.

On Wall Street, the Water Authority teamed with Poseidon in December 2012 to sell $734 million in construction bonds at 4.78 percent for the plant and 4.37 percent for the pipeline. In the end, it was the Water Authority's credit rating and history of successful high-profile projects that gave investors confidence in the project, which was one of the few large bond opportunities available to investors at the time. Because of near historic lows on the bond market, the final financing package would save $125 per acre-foot, or about $200 million over three decades, compared to projections used by the Water Authority board. Stonepeak, Poseidon's equity investor, posted $173 million in equity to complete the financing. When the Water Purchase Agreement was signed, the Water Authority estimated that Poseidon and Stonepeak could achieve an actual equity return of 10–13 percent, at the low end of the market range for a comparable investment. Third-party analysis after closing by Stratecon, Inc., pegged the equity return at about 8 percent in a generally favorable review of the landmark deal. "Since San Diego 'pays for performance,' the private capital transaction is based on the company earning returns from the contract," according to a *Journal of Water* analysis in January 2015 by Rodney T. Smith of Stratecon. "Rather than a detriment, the 'singular focus' of the private sector on earning returns is beneficial."[15]

The Carlsbad financing created an immediate buzz from coast to coast. It was named the Far West Deal of the Year for 2013 by the *Bond Buyer* magazine, which for more than a decade has selected outstanding municipal bond transactions for special recognition. "The deal—executed as a public-private partnership with Poseidon Resources—represents the first-ever project financing of a seawater desalination plant in the municipal market, establishing a new asset class for investors," the *Bond Buyer* said.[16] The partnership was also named the 2013 North American Water Deal of the Year by *Project Finance*, an international trade publication that annually highlights major industry accomplishments around the world.

An Aquatic Lifeline

Construction at the Carlsbad site officially started in the waning days of 2012, though it took a few months for activity to hit fever pitch, with hundreds of pipefitters, electricians, plumbers, welders, and other tradesmen working beneath towering cranes just beyond the sight of visitors to Carlsbad State Beach.

All told, the project would generate 2,500 jobs and infuse $350 million into the local economy during construction. Kiewit Shea Desalination, a joint venture of Kiewit Infrastructure West Co. and J. F. Shea Construction, designed and built the plant and the pipeline. Drought conditions allowed work to continue virtually unrestrained by rain delays through 2013 and 2014, pushing the project well ahead of its initial schedule. In December 2015 the Claude "Bud" Lewis Carlsbad Desalination

FIGURE 3-4

Carlsbad Desalination Plant: Process Flow Diagram

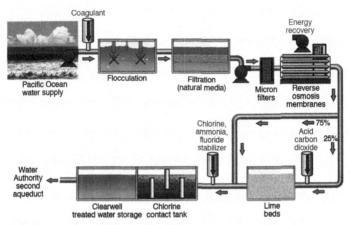

Source: San Diego County Water Authority.

Plant was dedicated and commercial water production began.[17] Four months later the plant was honored with a Global Water Award as the Desalination Plant of the Year for 2016 by Global Water Intelligence, publisher of periodicals for the international water industry. The award, announced at the Global Water Summit in Abu Dhabi, United Arab Emirates, went to "the desalination plant, commissioned during 2015, that represents the most impressive technical or ecologically sustainable achievement in the industry."[18]

Poseidon's six-acre site of industrial-zoned land sits on the edge of Agua Hedionda Lagoon, which is separated from the ocean by the Pacific Coast Highway. An inlet and an outlet cut under the roadway allow ocean water to circulate into the lagoon, where approximately 300 mgd are drawn into the existing power plant cooling water infrastructure.[19] One hundred mgd are diverted to the desalination plant, where the water is forced through RO membranes designed to remove the salts. Half of that water becomes concentrate, which is diluted with the remaining 200 mgd on its way back to the ocean. The other half of the water that enters the desalination plant—approximately 50 mgd—becomes pure freshwater, equal in volume to almost one-third of all the locally produced water in San Diego County.

It is pumped uphill ten miles to the east through the new conveyance pipeline installed underground, mostly under existing roads and public rights of way. The pipe intersects with the Water Authority's aqueduct, which was relined for about five miles to withstand the increased pressure exerted by the pumps; it had been a gravity-flow system until this new coastal water supply demanded reengineering the decades-old pipe. Figure 3-4 illustrates how the desalination process works. The desalinated seawater is stored and blended with treated imported water supplies at the Water Authority's Twin Oaks Valley Water Treatment Plant, then circulated through the regional distribution system—an aquatic lifeline for San Diego County more than two decades in the making.

Notes

1. "The Mission Period, from 1769 to 1834, was characterized by the efforts of the Franciscan missionaries to obtain water for the Mission San Diego de Alcala and for the Presidio." Mike Sholders, "Water Supply Development in San Diego and a Review of Related Outstanding Projects," *Journal of San Diego History* 48, no. 1 (Winter 2002)(http://www.sandiegohistory.org/journal/2002/january/sholders/).

2. "It has rained 12 of the past 29 days, depositing nearly 7 inches of water and making this San Diego's second wettest March on record": Amy Wallace, "Miracle March Dissolves Criticism of S. D. Mayor," *Los Angeles Times*, March 30, 1991.

3. Amy Vickers and David Bracciano, "Low-Volume Plumbing Fixtures Achieve Water Savings," *Opflow*, July 2014, pp. 8–9.

4. The California Senate Bill X7-7 of 2009 required water suppliers to increase water-use efficiency (www.water.ca.gov/wateruseefficiency/sb7/).

5. Large-scale indirect potable reuse started in California in 1962, when groundwater in Los Angeles County was recharged with treated wastewater as part of the Montebello Forebay Project.

6. In October 2013, Governor Jerry Brown signed the Water Authority-sponsored Senate Bill 322. "California needs more high quality water and recycling is key to getting there," the governor said in his signing message (https://www.gov.ca.gov/docs/SB_322_2013_Signing_Message.pdf).

7. Caitlin Cunningham, "The World of Desalination," editorial, *Membrane Technology*, November 2011, p. 3 (www.wwdmag.com/sites/default/files/03_Editorial TOC.pdf).

8. Quoted in Hari J. Krishna, "Introduction to Desalination Technologies," Texas Water Development Board Report 363 (December 2004) (www.twdb.texas.gov/publications/reports/numbered_reports/doc/R363/C1.pdf).

9. "Protecting GITMO's 'Achilles' Heel': Meeting Water and Energy Needs for 40+ Years at the Guantanamo Bay Naval Station," *Defense Systems Journal*, November 8, 2011 (www.dsjournal.com/greenatgtmo.html).

10. Michael D. Dettinger, Daniel R. Cayan, Noah Knowles, Anthony Westerling, and Mary K. Tyree, "Recent Projections of 21st-Century Climate Change and Watershed Responses in the Sierra Nevada," USDA Forest Service General Technical Report PSW-GTR-193 (2004), pp. 43–46, at p. 46. (www.fs.fed.us/psw/publications/documents/psw_gtr193/psw_gtr193_1a_04_Dettinger_others.pdf).

11. Tim P. Barnett and David W. Pierce, "When Will Lake Mead Go Dry?," *Water Resources Research* 44 (March 29, 2008) (http://onlinelibrary.wiley.com/enhanced/doi/10.1029/2007WR006704/).

12. The Water Purchase Agreement is available at the SDCWA website (carlsbaddesal.sdcwa.org/) under the Financial Affordability tab.

13. Two Water Authority member agencies, Vallecitos Water District and Carlsbad Municipal Water District, agreed to purchase a combined total of 6,000 acre-feet of desalinated water as their own local supply under separate agreements with the Water Authority. Those purchases effectively reduce the Water Authority's obligation to 42,000 acre-feet annually.

14. As a result of continuing improvements in the efficiency of RO membranes, the Carlsbad plant can produce 56,000 acre-feet per year without using all of the plant's RO capacity. In August 2016, the Water Authority's board certified a final Supplemental Environmental Impact Report that considered a potential future increase in the average annual plant capacity from 50 mgd to 55 mgd, which would require additional approvals.

15. Rodney T. Smith, "Economic Perspectives on San Diego County Water Authority's Carlsbad Desalination Project," *Journal of Water* (online), January 15, 2015 (http://journalofwater.com/jow/january-2015/).

16. "The Bond Buyer Announces Deal of the Year Finalists," *The Bond Buyer*, November 8, 2013 (www.bondbuyer.com/issues/122_217/the-bond-buyer-announces-deal-of-the-year-finalists-1057214-1.html).

17. The late Claude "Bud" Lewis was a longtime mayor of Carlsbad and a fervent supporter of the desalination project. Without his persistence and vision, the plant may never have been built.

18. See the website for the Global Water Awards (www.globalwaterawards.com/2016-winners#DesalinationPlantoftheYear).

19. The Encina Power Station is scheduled to shut down by the end of 2017. The existing intake system will be modified for desalination plant "stand-alone" operations and upgraded to comply with state regulations and permitting requirements.

Chapter 4

The Colorado River Story

JIM LOCHHEAD AND PAT MULROY

The use and management of Colorado River water are ruled by a series of negotiations, agreements, and court cases collectively known as the "Law of the River"—a body of law established with an imbalance that tells the ultimate story of deterministic planning.

The foundation of the law of the river is the Colorado River Compact of 1922, which allocated more water than has turned out to exist in the river. The compact has since been implemented and interpreted through a complex series of litigation, federal legislation, a treaty with Mexico, and federal agency operational criteria (see figure 4-1).

The river's many stakeholders recognized this critical imbalance early on, but they did not acknowledge it publicly until federal pressure, changing conditions, and controversies gave them no other choice. Their interactions began as an aggressive, winner-take-all fight, complete with lawyers, politics, and money.

Despite the inherent setup for dispute, the seven basin states—Arizona, California, Colorado, New Mexico, Nevada, Utah, and Wyoming— eventually began to see they had to work together to prevent unwanted federal intervention so they could retain the power to address their own

FIGURE 4-1

Average Flow of Major U.S. Waterways

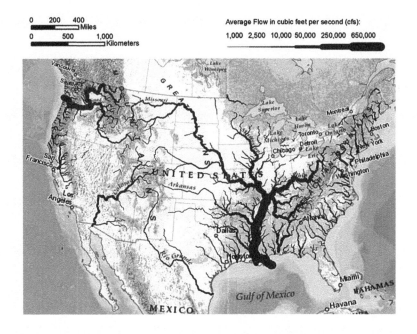

needs on home ground, with others who had to live with the results. As pressure rose, along with evolving conditions and the eventual reality of climate change, the states, municipal purveyors, and other river users worked to negotiate—rather than lose the time and money to litigate—their way through critical issues.

Stakeholders were forced to find ways to work from a watershed planning perspective, allocating risk not geographically but by sector—yet still within the Law of the River. Stakeholders managed to work within the framework of the Law of the River—its economics, politics, and social conflict—to build a water management and stewardship approach acknowledging not just the needs, rights, and interests of the seven states but also those of Mexico, Native American tribes, the federal government and the environmental community (see box 4-1 for a timeline).

This chapter provides the context for the laws and issues that surround Colorado River management and an overview of the steps state

BOX 4-1
Law of the River Timeline

1922	Colorado River Compact
1928–29	Boulder Canyon Project Act
1944	Mexican Water Treaty
1948	Upper Colorado River Basin Compact
1956	Colorado River Storage Project Act
1968	Colorado River Basin Project Act
2001	Interim Surplus Guidelines
2007	Interim Shortage Guidelines
2012	Minute 319 of the Mexican Treaty

and federal partners have taken in response to current and projected future conditions.

The Law of the River Gets Its Start

The Colorado River Compact, negotiated in 1922 and approved by Congress in 1928, was based on hydrology from a then recent, limited period that, in retrospect, was wet, with particularly abundant flows.

As a result, the compact's framers divvied up more water than nature could provide during sustained periods. They allocated, on a per year basis, 7.5 million acre-feet to the Upper Basin and 8.5 million acre-feet to the Lower Basin (an acre-foot is enough to supply an average of four households with water for a year). They even assumed additional water would be available for future allocation to Mexico, and that there would be surplus water for the states to divvy up later. The Upper Division states (Colorado, New Mexico, Utah, and Wyoming) and Lower Division states (Arizona, California, and Nevada) divided the basin at a point where the river's path hits Lee Ferry in Arizona, about 30 miles south of the Utah-Arizona border and just downstream from Glen Canyon Dam (see figure 4-2).

FIGURE 4-2

Colorado River Water Area Shared by Arizona, Colorado, New Mexico, and Utah

The compact negotiators focused on staking a claim to perpetual water supplies rather than relying exclusively on a sweeping prior appropriation, the "first in time, first in right" priority structure that would have rewarded the earliest water users and fueled a development race between basins. Although important as the basins grew, the allocation structure certainly would never have been enough to prevent differences and conflicts between the basins.

Lower Basin representatives wanted support for reservoir storage to help provide reliability in dry times and protection from floods in wet times. California, for one, wanted flood protection after the 1905–07 conditions that contributed to the creation of the Salton Sea in the desert. The Lower Basin also wanted an "All-American Canal" to sustain Imperial Valley irrigation, which was to replace the existing Alamo Canal, routed through Mexico. The 1928 Boulder Canyon Project Act approved the Colorado River Compact and gave the basin the go-ahead for the All-American Canal System, the Imperial Diversion Dam, and what would become Lake Mead's Hoover Dam. Ultimately, though, the Lower Basin would struggle with what would be called its "structural deficit," or the inability to fill storage at a rate adequate to sustain use.

The Upper Basin states focused on securing and protecting a defined right to a reliable water supply for their use and for future development needs. Like the downstream states, the Upper Basin states prioritized storage and wanted the ability to develop reservoirs with little to no litigation or federal intervention. Storage would become even more important in the face of forecasted deficiencies, a burden known as the "hydrological leftovers" the compact placed on the Upper Basin through a requirement to not cause the flow at Lee Ferry to drop below a total of 75 million acre-feet in any ten-year running period. The compact left the Upper Basin with no fixed allocation, only what nature would provide minus the fixed obligation to the Lower Basin.

Before developing storage, the Upper Basin states had to agree to individual states' allocations, which they did in the 1948 Upper Colorado River Basin Compact. After witnessing the problems that came out of the compact's hard numbers, and with the uncertainty of the hydrological leftovers problem, the Upper Basin's compact divvied up water in percentages of flow. In 1956 the Colorado River Storage Project Act authorized what would become the Flaming Gorge, Navajo, and Glen Canyon Dams. Glen Canyon would house Lake Powell, to interplay with the Lower Basin's similarly sized Lake Mead. Hydropower revenues would become key, as they would help repay the costs of dam construction.

Early discussions at both the federal and basin levels addressed the river users' need for help in financing and developing major system storage and hydropower projects. The basins would ultimately benefit from compact enforcement at the federal level, including the secretary of the interior's responsibility to operate the reservoirs that supported compact deliveries.

However, the states also wanted to limit federal involvement that could impair their ability to make decisions at their own level, among the people who had to live with the decisions. The Colorado River Compact and other early agreements and proposals got the seven states together to start discussing their needs and concerns. The negotiations were difficult and frustrating, but they offered a starting point. For example, early agreements addressed two basic yet critical issues: restricting the Upper Basin from withholding water and restricting the Lower Basin from ordering water it could not reasonably use.

These discussions would not be enough to avoid the heated conflicts that would arise, along with the evidence that demands were growing and supplies were lacking. In addition to overabundant hydrology, Colorado River Compact allocations were based on narrowly anticipated demand, which was driven almost exclusively by irrigated agriculture. The compact framers could not anticipate future urban demands both from within the basin and outside it. Los Angeles and San Diego, Phoenix, Las Vegas, Denver, Salt Lake City, and Albuquerque all depend in whole or in part on Colorado River water to support some 30 million people and $1.4 trillion per year in economic output. Nor could the framers have anticipated the environmental values (not to mention laws) and recreational economies that depend on the river's flow.

Arizona-California Conflict

No sequence of events more clearly illustrates how imbalanced Colorado River allocations fueled disputes than the historic conflict between Arizona and California over water rights.

Starting in 1931, multiple rounds of litigation related to water distribution between the two states went before the U.S. Supreme Court. Arizona—which, like the Upper Basin states, was concerned about the rapid rate of development in California—ultimately prevailed in a 1963 decision that limited California to its basic 4.4 million acre-foot annual allocation.

When the court ruled in favor of Arizona, it ignored tributary uses (particularly on the Gila River) and allocated 7.5 million acre-feet per year out of Lake Mead for use in the Lower Basin. However, the court failed to account for losses of water from reservoir evaporation and river transit losses, which have averaged some 1.2 million acre-feet annually. (The current low levels in Lake Mead reduce evaporation by as much as 200,000 acre-feet compared to average water levels.) Also, the 1944 treaty with Mexico allocated 1.5 million acre-feet per year to that country.

Thus, total demands from Lake Mead exceed 10 million acre-feet annually. However, the Upper Basin is obligated to supply an average of only 7.5 million acre-feet per year, plus potentially half of the Mexican allocation in times of shortage depending on legal interpretation of the compact for a maximum total annual average of only 8.25 million acre-feet. In contrast, the Upper Basin's percentage-based compact allowed for an aggregate use of less than the 7.5 million acre-feet per year, which could account for the realities of system loss and evaporation. The Law of the River had created a structural deficit problem of water supply in the Lower Basin—too many uses and not enough water.

The Lower Basin's structural deficit became even more complicated when California exacted its retribution for the loss in *Arizona v. California* through raw politics. In 1968, federal legislation authorized the Central Arizona Project (CAP), a canal system that would bring a portion of Arizona's Colorado River allocation to the burgeoning metropolitan areas of Phoenix and Tucson, and for extensive irrigation. California's large congressional delegation forced the subordination of Arizona's CAP deliveries to assure California it would continue to receive its full 4.4 million acre-feet each year, even in a shortage. In other words, the primary water supply for Phoenix and Tucson could run dry before California's farmers or cities would lose a drop.

Arizona knew at the time that the CAP's low-priority status on the already overallocated Colorado River system curbed its potential as a reliable supply. Tensions grew by the early 1990s, when agriculture used less CAP water than Arizona had anticipated, making unused water available for California. California was using up to 5.3 million acre-feet per year. Arizona was concerned that it could not yet demonstrate beneficial use of its available water supply and had to let it continue to California, supporting that state's reliance on essentially temporary supplies.

The CAP would suffer in times of shortage, a situation made certain by the Lower Basin's structural deficit. And climate variability would only make matters worse. These issues made it increasingly clear that, without evolving, the Law of the River couldn't adequately or equitably address this looming crisis. But the states could not keep turning to litigation, federal intervention, or structural projects. These old measures could not promise to be feasible, timely, or even appropriate in the face of important issues. The states needed to move toward increased cooperation, compromise, innovation, and action. Their efforts often raised difficulties and frustrations but ultimately altered management on the Colorado River. One early example is the creation of interstate banking efforts.

Interstate Banking

Arizona looked for other ways to bolster its use of water: water banking. In 1996 the Arizona Water Banking Authority was formed, providing a mechanism for the state to bank or store its unused Colorado River apportionment in Arizona's underground aquifers for future use.[1] This arrangement addressed the management of both surface and groundwater resources, which allowed Arizona to address declining groundwater levels as well as mitigate Colorado River shortages, an effort achieved by saving water during wet years for use during dry years.

The state legislation that created the authority also formed an interstate bank to give Nevada and California the opportunity to bank water in Arizona. Federal regulations to facilitate interstate banking were subsequently approved in 1999 and 2001, prompting Arizona and Nevada to begin formal negotiations for Nevada's participation.[2] The water goes toward municipal, industrial, and agricultural uses, as well as tribal rights.

These early efforts to establish, clarify, and expand business arrangements for interstate banking efforts helped pave the way for other banking projects in the Lower Basin states. These collaborations toward adaptive capacity, which have afforded a new level of flexibility within the Law of the River's existing framework, help Lower Basin water users meet their needs within their own basin. Conceptually, these programs benefit the Upper Basin by allowing the Lower Basin to begin to address its structural deficit problem and meet its demands from within the Lower Basin.

The shift from fighting to collaborating gave the states new opportunities to address critical issues beyond what the Law of the River had anticipated. And their ability to do so, and to find ways to manage demand in response to changing hydrology, would become very important very soon.

2001 Interim Surplus Guidelines

By 1990 it was apparent that Nevada, fueled by unprecedented urban growth, would soon require more water than its base allocation. Simultaneously, Arizona was embarking on an aggressive banking strategy to protect future urban development in the Phoenix and Tucson areas.

And California was using more than its basic allocation. A buffer had always existed in the form of surplus declarations, which in favorable hydrological conditions—exceeding what was needed for normal operational pumping and release from Lake Mead—prompted the secretary of the interior to authorize California uses above its basic allocation. In all but two years, from 1965 to 2001 California used some 800,000 acre-feet more than its 4.4 million acre-foot annual base allocation. California had become dependent on these additional supplies. Moreover, because of their junior priority within California, urban uses on the coastal plain from Ventura to San Diego would have to be reduced if California was limited to its base allocation all at once.

In the early 1990s a record drought hit California, and the state asked the secretary of the interior for continued surplus Colorado River water. But the Colorado River was also in drought, and the other states in the basin opposed the request.

Secretary of the Interior Bruce Babbitt brought the states together in the mid-1990s to begin addressing surplus-related issues. The states knew they had to fix the problem among themselves or else be prepared to face a decision handed down from the federal level. The states' decades-long history of contention strained early surplus negotiations, but the conditions also provided a critical, timely reason for the states to work toward mutually agreeable guidelines for system operations and supply management in times of plenty.

In 2001, as his final act in office, and after nearly 10 years of negotiations among the states, Department of the Interior, and major water users in the basin, Secretary Babbitt signed the Colorado River Interim Surplus Guidelines, reflecting decisions based largely on the agreements reached among the seven basin states.[3] This landmark set of guidelines created, for a time, a new form of surplus tied to specific elevations in Lake Mead. This water could be used beyond a state's base allocation through the year 2026, giving water planners the increased predictability of reservoir conditions to drive decisions. The guidelines also established a process through which California could reduce its use without jarring effects to its economy, thereby helping achieve a soft landing to its basic allocation of water.

Quantification Settlement Agreement

The agreement establishing the Interim Surplus Guidelines was predicated on the understanding that California would, within fifteen years, reduce its annual Colorado River withdrawals to 4.4 million acre-feet. In 2001 Secretary Babbitt's successor, Secretary Gale Norton, threatened federal suspension of the Interim Surplus Guidelines by mandating that a plan to realize this reduction, the Quantification Settlement Agreement (QSA), be signed before 2002 came to a close.

One of the more than forty agreements necessary for the QSA to come to fruition related to a transfer of agricultural water from the Imperial Irrigation District to urban uses by the San Diego County Water Authority. In addition to financial and water resource issues, concerns related to the agreement's potential impact on migratory birds, fishes, and invertebrates sustained by the Salton Sea played a role in the proposed accord—an early nod to the importance of managing the river from a wider watershed perspective, but foregrounding a problem that continues to fester unresolved to this day.

California failed to meet its QSA deadline, and the Interim Surplus Guidelines were temporarily suspended until the California agencies reached agreement, although interests in the Imperial Valley immediately entered litigation challenging the QSA. Either way, California reduced its annual use from 5.2 million acre-feet to 4.4 million acre-feet, but not on the soft landing plan: drought came after the surplus, and California was forced to reduce its use immediately.

As fate would have it, the surplus began to dry up as quickly as the ink of the signatures on the 2001 Surplus Guidelines. As if on cue, the Rocky Mountain snowpack, which feeds the Colorado River, decreased dramatically, and Lake Powell and Lake Mead began to recede as if someone had opened a drain. By 2003 Lake Mead's elevation had dropped approximately seventy-five vertical feet, a loss of trillions of gallons of storage in just three years. Lake Powell took a similar hit. Hydrological variability and risk heightened tension between the basins.

2007 Shortage Guidelines

As the drought took hold, the states that had spent years divvying up the spoils of surplus found themselves staring at potential shortages with no rules to guide them. The timing and amounts of cutbacks in the Lower Basin were set to be determined by the secretary of the interior each year. But the Law of the River established no specific criteria for the secretary to follow, leaving water managers with limited information about future operations, and therefore little or no certainty for planning at low reservoir elevations.

As dry conditions gripped the region, Arizona insisted Lake Powell had the supplies to continue federal government releases of at least 8.23 million acre-feet a year to the Lower Basin. This amount, a minimum objective release in the operating criteria since 1970, relates to the Lower Basin's Colorado River Compact allocation and the 1.5 million acre-foot entitlement for Mexico. The Upper Basin disputed this number, as it implies a requirement to supply half of the obligation established in the 1944 Mexican Treaty, even in times of Lower Basin surplus. As Arizona stood by the 8.23 million acre-foot volume, Colorado established a $10 million fund for a potential onslaught of litigation and sought to limit the risk to Upper Basin obligations.

In 2005, Secretary of the Interior Gale Norton sent a letter to the states exerting the authority to determine release operations for Lakes Mead and Powell under shortage conditions, including a possible reduction in the 8.23 million acre-foot minimum objective release to the Lower Basin. Norton gave the states two years to make their own decision, or else she would make it for them.

The states were able to reach an agreement in 2007, an effort that supported the federal 2007 Record of Decision for Colorado River Interim Guidelines for Lower Basin Shortages and Coordinated Operations for Lake Powell and Lake Mead (aka Interim Shortage Guidelines), which recognized the connections between the two reservoirs and provided for closer balancing of storage between them.[4]

The Interim Shortage Guidelines also defined cutbacks for water users based on the elevation of Lake Mead, with the intent of protecting the lake's elevation by reducing outflows if drought conditions worsened. The guidelines provided for the release of less than 8.23 million acre-feet from Lake Powell. This helped relieve the Upper Basin of the fear of creating the precedent of always having to provide half of the U.S. obligation to Mexico.

The decision also implemented an Intentionally Created Surplus (ICS) program, which gave Lower Basin users the ability to store water in Lake Mead through extraordinary conservation, imports, and other innovations. The decision addressed ICS storage limits and releases from Lake Mead in normal conditions as well as times of surplus and shortage.

The shortage-sharing agreement set a precedent of cooperation among Colorado River users and fostered an attitude of cooperation before litigation, all while continuing to take advantage of the secretary of the interior's role as contracting officer but not intervener. It added flexibility to system operations, which helped protect Upper Basin storage and demonstrated the need for Lower Basin states to address their issues in times of shortage.

Minute 319 to the Mexican Water Treaty

The Interim Shortage Guidelines were heavily influenced by discussions being conducted across international borders around the same time. Mexican officials were protesting the documented U.S. government assumption that Mexico would, like the states, take a percentage reduction in times of shortage. Mexico also was displeased with the drying up of wetlands by QSA-required work to line the All-American Canal. These issues, along with environmental concerns, spurred both dispute and discussion.

Even the states chose to join the discussions after they agreed to the 2007 Interim Shortage Guidelines. The states jointly requested that the federal government commence discussions with Mexico to bring the nation into the shortage management structure. The water provisions of the Mexican Treaty included the condition that Mexico could be shorted in "extraordinary drought"—a concept that has not been defined to this day. The states wanted Mexico to share in the burden and not receive water under shortage conditions.

Signed in 2012, Minute 319 to the Mexican Treaty brought Mexico into the partnership of managing the Colorado River for the next five years. The nation agreed to reductions in delivery during shortages through the year 2017, using the same elevations that trigger shortages in the United States. On the other hand, Mexico was for the first time allowed water beyond its treaty share during favorable reservoir conditions. Minute 319 granted Mexico the ability to store in the United States water that it conserved in Mexico as a buffer against shortages.

Although Minute 319 represented landmark agreements between nations, the states' participation set a standard. As water users, the seven states had and continue to have a huge stake in Colorado River issues, and decided to make seats for themselves at the Minute 319 negotiation table. The result was increased success, demonstrating that partnerships must be made at multiple stakeholder levels.

Even more, the true precedent Minute 319 set was the inclusion of new interests at the negotiation table—interests that also would become integral to the Interim Shortage Guidelines. For the first time, major Colorado River discussions included nongovernmental organizations (NGOs), which shed light and credibility, in the forms of scientific data and financing, on the river's complex relationship with the environment. The presence of major conservation organizations resulted in important provisions in Minute 319, including commitments to provide freshwater flows and to conduct ecological restoration programs in the Colorado River Delta. The two nations also agreed to fund projects that would support water conservation efforts and the provision of environmental flows in Mexico. And the NGOs, which pushed the concept of conservation before shortage, offered important ideas that were incorporated into the Interim Shortage Guidelines.

More Efforts: Seeking Solutions, Setting Principles

As water managers, the states recognized they had a responsibility to consider the implications of all intervening variables, including climate change. In 2012 they engaged in the comprehensive Colorado River Basin Water Supply and Demand Study to quantify current and projected supply-and-demand imbalances in the system over the next half century, incorporating for the first time climate change projections.[5] This helped users understand how critical the river's situation was and that they needed to take an all-in approach—there was no time for individual, narrower disputes or concerns. Climate analysis reinforced critical supply-and-demand issues, underscoring the need for more immediate and proactive measures.

The study concluded that if no additional water management actions were undertaken, a wide range of future imbalances was plausible, primarily because of the uncertainty in future water supply. The best available science at the time calculated a median 9 percent overall reduction in natural (unaltered by humans) streamflow over the next fifty years and an imbalance in supply and demand of 3.2 million acre-feet by 2060 (based on a comparison of the median water supply projections against the median water demand projections). Modeling also projected an increase in both drought frequency and duration.

While the report provided single projections to simplify results, it is important to note that the same climate model showed more than a 9 percent decrease in natural streamflow, while other models showed an increase. The most important finding may be the expectation that warming will continue across the basin, increasing evaporation, plant consumption of water, and other water losses. Research in 2016 found that the ongoing drought, the worst on record, was being driven by warmer temperatures, leading to increased evaporation and other losses, rather than by a lack of precipitation, as was the case in previous droughts.[6] This finding could be particularly disturbing as a warning that the current drought might become the new normal.

The year 2013 turned out to be the worst water year since 2002, when the Colorado River received less than one-quarter of its historical runoff. There was one major difference: back then, system reservoirs were relatively full, but as of 2013, both were well below half of their combined capacity.

The Lower Basin states agreed to address their structural deficit, and the Upper Basin states began to work together to make their hydrological leftovers problem more secure. Logistically, any ground the Lower Basin could gain in its attempts to manage the levels of Lake Mead, and efforts in the Upper Basin to protect elevations in Lake Powell, should benefit both reservoirs, thanks to better operational synergy between them.

Federal and state entities came together to explore efforts to help slow the decline of levels in Lakes Mead and Powell, and started working with novel partners—municipalities—which in turn agreed to support the 2014 Pilot System Conservation Agreement, a program designed to increase the availability of Colorado River supplies through temporary, compensated, and voluntary conservation or water-use reductions. The collective goal was water to benefit the system as a whole rather than an individual basin, user, or entity.

Conclusion: Lessons Learned

The Colorado River's stakeholders initially sought to claim their stake by fighting for their piece of the pie. But the Law of the River's allocations, combined with extended droughts, set them up with a smaller pie than initially thought, and this critical imbalance would eventually give water users no choice but to sit down at the table and talk.

Water users learned they had to respect the separate issues as well as the interdependencies that exist within the entire Colorado River system—which must be managed as such, rather than as a series of small systems. Knowing this, the seven basin states started to take a watershed-level approach that considered the risks and gains faced by all water users, including urban, agricultural, environmental, and recreational users.

It took collaboration, which meant being honest yet patient, staying true to their own needs yet learning to understand the needs of others. Over time, negotiations helped water users find ways to adjust infrastructure plans, improve water efficiency, and take steps to improve the reliability of their deliveries.

But those efforts, although admirable, were just practice runs. The story is far from over. Although stakeholders made great strides to stop fighting to win and start talking to reach solutions, the looming concern of climate change adds an entirely new level of urgency. Climate change not only speeds up the imbalances that came out of the first agreement on the Law of the River, but also accelerates all of the challenges the states face now, including the need to have a critical conversation about living within their means.

The states must go far beyond their collaborations of the past, and they will have to do so together. No single entity or project—such as new large federal reservoirs or desalination projects on the coast—can feasibly offer an easy way out. They have no choice but to find solutions through addressing their own big problems, the Upper Basin's hydrological leftovers and the Lower Basin's structural deficit. As their populations continue to grow, the states will need to take past actions of cooperation to the next level of innovation.

There are plenty of challenges in sight, including California's unique priority to its 4.4 million acre-foot annual allocation. With this lack of incentive, will the six states be able to bring California to the table? California already houses its own looming concerns, the Salton Sea and the Bay-Delta area. Stakeholders may find the answers to systemwide problems sitting at the confluence of collaborations that have helped California find solutions.

Uncertainty has become an established element of the river's future that absolutely requires a water-efficient way of thinking. In the face of uncertainty and lurking imbalance, the states face the slow, grueling process of collaboration. But if water users can keep trying, if they keep being honest and committed, and keep meeting and bringing new ideas to the negotiation table, their collaborative work may continue to serve as a remarkable example of building adaptive capacity and finding flexibility within the rigidity of the regulatory and structural frameworks on the Colorado River.

When the framers of the Colorado River Compact divided the waters of this meager river, they assumed that major storage projects along the system would provide an ample buffer against the variability of flows. Not only did the science of the time blind them to what the real average flow of the system was, they never imagined the scope of future demands, or the devastation climate change could wreak on the system.

However, as flawed as some of their underlying assumptions were, there is still great wisdom in the document. It has proven to be highly flexible and pliable, adaptable to conditions as these unfold. It allows seven states to carve a common path forward without upsetting the fragile balance struck by the document. It defied the biblical underpinning of the western water law of prior appropriation and created seven equal partners, none with greater or lesser rights than the others. What came after it began to upset that balance as jockeying for a superior position among the partners played itself out. Much of that is now under the microscope, as the partners begin to realize the impractical nature of actually applying these advantages. All the agreements that have been forged over the past twenty-five years are in reality a journey back to the beginning, back to the concept that seven equal partners have to craft a common path forward if they are all going to survive.

Notes

1. Arizona Revised Statutes 45-2401 *et seq.*

2. Storage and Interstate Release Agreement pursuant to 43 C.F.R., Part 414, Offstream Storage of Colorado River Water and Development of and Release of Intentionally Created Unused Apportionment in the Lower Division States.

3. Department of the Interior, Bureau of Reclamation; Colorado River Interim Surplus Guidelines, 66 Fed. Reg. 17 (2001).

4. Department of the Interior, Bureau of Reclamation; Record of Decision for the Adoption of the Colorado River Interim Guidelines for Lower Basin Shortages and Coordinated Operations for Lake Powell and Lake Mead, 73 Fed. Reg. 19,873.

5. The study is available at the Bureau of Reclamation, Lower Colorado Region, website (http://usbr.gov/lc/region/programs/crbstudy/finalreport/index.html).

6. Brad Udall, Colorado State University, personal communication, 2016.

Chapter 5

Nebraska's Water Governance Framework

ANN BLEED

In the arid West, where competition for the use of limited water supplies is high, regulating water use will be paramount for the survival of cities. However, many suggest that the traditional top-down water governance framework typical of most western states is too rigid to cope with climate change. For more than forty years the state of Nebraska has relied on a unique, locally controlled water governance system to manage and regulate many of the state's natural resources, including groundwater. This chapter explains why this more flexible governance system could be one of Nebraska's best strategies for coping with the stresses and uncertainties of climate change.

Nebraska's experiment with a locally controlled groundwater governance system is not insignificant. Nebraska has more irrigated acres than any other state, and more than all but a dozen countries.[1] In 2010, water for agriculture accounted for 73 percent of the total water withdrawals for the state; withdrawals for public use accounted for only 3 percent.[2] Groundwater provides water for more than 83 percent of Nebraska's irrigated acres[3] and 96 percent of Nebraska's public water supplies.[4] Thus, Nebraska provides a significant case for studying a large-scale, locally controlled groundwater governance system.

The Water Governance Challenge

In the past, the general assumption was that environmental and social systems are predictable, and that changes in these systems are incremental and linear.[5] Accordingly, most of the legal and governance frameworks for natural resource management and regulation in general, and water governance in particular, have been based on the presumption of stability[6] and on assumptions that socioecological systems are predictable and changes in them are incremental and linear.[7] As a result, most water governance systems in the West are monocentric, top-down systems that focus on applying relatively inflexible rules to limit actions and water use.[8]

Will such systems have the resilience to adapt to climate change? "Resilience," in the words of Brian Walker and David Salt, "is the capacity of a system to absorb disturbance and still retain its basic function and structure."[9] For human societies, it is important that the socioecological systems on which society depends continue to function to provide the benefits humans need and, ideally, desire.[10] From resilience research, we now know that both environmental and socioecological systems are not linear and stable but rather are complex, multiscalar, interactive, and dynamic, and, when stressed, will produce sudden, unexpected, and sometimes unwanted results and administrative paths.[11]

Fortunately, based on a significant body of work by Elinor Ostrom, and her colleagues, we also know that monocentric, top-down water governance is not the only way to successfully govern the use of water resources. Ostrom's research described many cases of commonly held resources, such as water supplies, that were successfully governed at the local level for long periods of time, up to 1,000 years.[12] Using game theory and laboratory experiments, as well as the documentation of many examples of successful, locally controlled governance, Ostrom developed a list of eight guidelines for successful locally controlled natural resources governance systems.[13] Since her seminal publication, Ostrom and others have elaborated and expanded on this list.[14]

Nebraska's Water Governance and Legal Framework

As in many other western states, the Nebraska State Department of Natural Resources (state DNR) regulates surface water under the prior appropriation system. In a prior appropriation system, the date of the permit determines which appropriator has priority to water. By contrast, in Nebraska groundwater is regulated by twenty-three locally elected natural resources district (NRD) boards under a reasonable use/correlative rights system in which, in times of shortage, the resource is shared among users.

The local NRDs were created in 1972 to consolidate 154 local special-purpose districts into a more effective natural resources governance system.[15] Preservation of local control was, and still is, a high priority in Nebraska. The NRDs were given a wide range of authority over the state's natural resources, including prevention of erosion, floods, flood damages, and pollution; drainage improvement and channel rectification; soil conservation; solid waste disposal and sanitary drainage; forestry and range management; the development and management of fish and wildlife habitat and recreational and park facilities; and the development and management of water for beneficial uses.[16] It is important that the NRDs were also given the authority to levy their own taxes to fund their operations.[17]

When the NRDs were created, the state was primarily concerned with flood control, water drainage, and soil conservation. Thus the boundaries of the NRDs were drawn to coincide with river basin, surface watershed boundaries. To maintain a governance system focused on local control, the governance of some of the larger river basins was divided among as many as six or more NRDs, depending on how a basin is defined. It was not until 1975 that the NRDs were also given the primary authority to regulate groundwater.[18] In Nebraska, the groundwater aquifer boundaries rarely coincide with the surface watershed boundaries, so the NRDs not only must share the governance over certain watershed boundaries, they must also share the governance of certain groundwater aquifers.

For years, Nebraska's water laws did not recognize that surface water and groundwater could be hydrologically connected. Although the legislature took some early tentative steps toward integration,[19] it was not until 2004, in large part in response to conflicts between surface water and groundwater users,[20] that the legislature passed a comprehensive law integrating the management of surface water and groundwater.[21] This legislation was drafted by a forty-nine-member Water Policy Task Force representing a wide range of water interests across the entire state. Because of the importance of local control, the task force maintained the split between the governance of surface water by the state DNR under the prior appropriation system and the governance of groundwater by the NRDs under the modified correlative rights system.[22] However, in areas where surface water and groundwater were hydrologically connected and were fully appropriated, that is, where the state DNR determines there is no excess water available for new consumptive uses, the new law requires that the affected NRD and the state DNR jointly develop and implement an integrated management plan. State law requires that such plans ensure that their implementation will "sustain a balance between water uses and water supplies so that the economic viability, social and environmental health, safety, and welfare of the river basin . . . can be achieved and maintained for both the near term and the long term."[23] The integrated management plan must also ensure that the water supplies for surface water appropriators and groundwater-well owners existing at the time the state designates the basin as fully appropriated will not be depleted by additional water uses.[24] In basins that were already overappropriated, that is, where water uses already exceeded the supply, the law requires the development of a basinwide plan that, over time, shall reduce water uses to the point where the basin is no longer overappropriated.[25] Nine NRDs have completed required integrated management plans, and all but one NRD have completed or are currently working with DNR to develop voluntary integrated management plans.[26]

Can the NRD Governance System Be a Resilient Alternative?

Based on Ostrom's and others' research, my colleague, Christina Babbitt, and I developed a list of seventeen criteria that are indicative of resilient, locally controlled governance systems and used these criteria to try to determine whether Nebraska's NRD governance system is likely to succeed in maintaining the state's water supplies.[27] What follows is a description of Nebraska's water supplies and uses, and a synopsis of our assessment of the resilience of Nebraska's water governance system.

Nebraska's Water Supplies and Uses

Nebraska is blessed with ample water supplies. The High Plains Aquifer (frequently called the Ogallala Aquifer)[28] extends from Wyoming and South Dakota through Kansas and Oklahoma and into Texas. In Nebraska the High Plains Aquifer, which covers 84 percent of the state, ranges from 100 feet to 1,000 feet of saturated thickness.[29] The total water in storage in the High Plains Aquifer in the year 2000 was about 2,980 million acre-feet, 2,000 million acre-feet of which are under Nebraska.[30]

Even though Nebraska is blessed with abundant water supplies, and the importation of surface water from other states has recharged the groundwater, contributing to groundwater table rises of up to 120 feet, the use of groundwater primarily for irrigation has caused groundwater table declines in many areas of the state, with declines of as much as 60 to 120 feet in some areas.[31] Moreover, in many areas of the state cities and towns experience not only water quantity problems in dry years, when pumping for irrigation is high, but also water quality problems, primarily related to high levels of nitrates in the groundwater. The use of nitrogen in fertilizer, which then becomes a non-point-source pollutant, is a major source of groundwater pollution in Nebraska. Although the federal Clean Water Act has provisions to prevent non-point-source pollution, the U.S. Environmental Protection Agency has found it difficult to adequately curtail this source of pollution.[32] Thus, many cities and towns are already struggling to find additional good-quality water supplies. Looking to the future, the city of Lincoln, the state's capital, is contemplating building a pipeline to the Missouri River at an estimated cost of $500 million (2013 dollars) to provide water for the future,[33] but for most of the state's municipalities, the Missouri River is too far away to be a viable source of water. Therefore, even in Nebraska, water governance is critical to maintaining sufficient good-quality water supplies for its cities and towns.

The Impact of Climate Change on Nebraska

According to a report by the University of Nebraska, in Nebraska climate change is expected to bring warmer temperatures and more extreme climatic events. By the end of the twenty-first century, temperatures are expected to rise by 4° to 9° F, and the number of high-temperature stress days, with temperatures over 100° F, is also expected to increase. Precipitation is expected to decrease, particularly in western Nebraska, and the intensity and frequency of droughts are expected to increase. The higher temperatures will also cause a change in the amount of snowpack in the mountains, which will have an adverse impact on Nebraska's ability to access and store runoff. Soil moisture is expected to decrease, which will lead to increasing demands for water for irrigation. Irrigation demands will compete with municipal demands.[34]

Evaluation of the NRD Governance System

The criteria listed in table 5-1 were used by Bleed and Babbitt to evaluate the resilience of the NRD governance system.[35] For this evaluation, each criterion, and whether or not the NRD governance system meets that criterion, will be briefly described.

To successfully govern water resources, there must be a formal basin-wide governance structure with a high level of authority.[36] A holistic approach to water and natural resource management is also important because water resources are affected by many activities that extend beyond how the water resource itself is used. For example, increased pavements in cities, soil erosion, and even air pollution have adverse impacts on both the quantity and quality of a basin's water supply. Conversely, soil conservation activities, a decrease in the use of farm chemicals, and the development

TABLE 5-1

Criteria for Assessing Successful Water Governance

Criteria

1* River basin approach—The governance system should be focused on the ability to holistically manage a basin's water system, as well as on other key aspects of the basin's ecosystems.

2 Clearly defined boundaries—Both the individuals who have rights to withdraw from the resource and the boundaries of the resource must be clearly defined.

3* Recognition of rights to organize at the local level (local control)—The rights of users to devise their own institutions are not challenged by external governmental authorities.

4 Rules to prevent overharvesting—Rules should be in place to restrict use to prevent depletion of the resource. The purpose of these rules is not to allocate water among uses or to water users.

5 Graduated sanctions—Users who violate rules are likely to receive graduated sanctions dependent on the seriousness and context of the offense.

6 Congruence between appropriation/provision rules and local conditions; proportional equivalence between benefits and costs—A one-size approach to water governance does not fit all situations; the approach must reflect the conditions of a given locale and must provide benefits and costs acceptable to water users.

6* Monitoring—Monitors, who actively audit biophysical conditions and user behavior, are at least partially accountable to the users, or are the users themselves.

7* Adequate funding—A stable and sufficient funding source is essential to developing and sustaining water management programs.

9 Secure tenure rights—To encourage sustainable practices and investment, water users have assurance that their right to the resource is secure for the long term.

(continued)

10 Rapid access to low-cost, effective conflict resolution mechanisms—Users and their officials have rapid access to low-cost, local arenas to resolve conflict among users or between users and officials

11* Effective and efficient communication systems—Effective and efficient communication must be in place; groups that do not communicate well are more likely to overuse the resource.

12* Leadership—Good leadership involves making difficult choices that are in the best interest of society as a whole, providing overarching direction to constituents, and being willing to be a part of the long-term decisionmaking process.

13* Trust—Trust is an essential component in building reciprocity and cooperation.

14* Collective-choice arrangements—Ability to Influence Rules and Collaboration—Most individuals affected by harvesting and protection rules are included in the group which can modify these rules.

15 Equity and procedural fairness—Mechanisms are available to achieve equity and procedural fairness. Despite differences in how people use and value water, it is essential that all water users feel they are treated fairly.

16* Adaptive management—Water institutions must be able to adapt to changing conditions. To adapt, they must have the freedom and flexibility to develop and implement innovative solutions, learn from new information, and revise their action plans.

17* Nested enterprises, polycentric governance, and adaptive comanagement—Local institutions are part of a larger, integrated network with different hierarchies and scales that collaborate with each other to manage the resource.

Source: Adapted from Ann Bleed and Christina Hoffman Babbitt, Nebraska's Natural Resources Districts: An Assessment of a Large-Scale Locally Controlled Water Governance Framework, Policy Report of the Robert B. Daugherty Water for Food Institute (University of Nebraska, 2015).

Note: Asterisks indicate the criteria also thought to be necessary to provide the adaptive capacity necessary to sustain water supplies given the challenges of climate change.

of wetlands and recharge basins, which also may provide wildlife habitat areas, can prevent water pollution and flooding and increase recharge to groundwater reservoirs. A river basin approach is particularly important in water-scarce basins, where demands for water and the impacts of change are high.[37] The NRD legal framework established the boundaries of the NRDs to coincide with river basin boundaries and also gave the NRDs authority over a wide range of natural resources. The NRDs have used their authority not only to regulate water but also for more holistic resource management. Across the state, NRDs are regulating the use of nitrogen fertilizers that contribute to non-point-source pollution. Federal law has not been very successful in regulating non-point-source pollution, but nitrate pollution has been reduced as a result of NRD actions.[38] NRDs have also cooperated with cities to develop rain gardens and other natural areas to decrease water contamination and flooding and increase groundwater recharge.[39]

The boundaries of the resource must clearly delineate who has the right to withdraw water. Without defining the boundaries and closing the use of the resource to outsiders, local appropriators face the risk that any benefits they contribute to the effort will not return to them.[40] Although the NRD boundaries cross both surface watershed and groundwater aquifer boundaries, there is no question of which NRD controls the use of groundwater in any given area.

According to Ostrom, the ability to establish local rules has allowed the evolution of fairly complex rules that, unlike rules established by external government officials, are accepted and enforced by the stakeholders without external government authority.[41] On the other hand, when external governmental officials do not understand the local system but presume that only they have the authority to set the rules, systems previously robust for long periods of time have largely been destroyed.[42]

Ostrom also states that without limiting use to prevent overharvesting, local appropriators face the risk that those who have made investments based on the availability of the resource will not receive as high a return as expected on their investment. Moreover, if there are a lot of appropriators and a high demand for the resource, the chances the resource will be overused are also high.[43]

The legal framework for the NRD system not only recognizes but strongly supports the concept of local control and the right to organize at the local level. Furthermore, under this local-control framework, the NRDs have developed rules to try to prevent overharvesting.[44] Because the local NRD, not the state, developed and enforced the regulations, many new management and regulatory actions in Nebraska have been accepted, enforced, and successfully implemented without the need for many sanctions.[45] For example, the Upper Republican NRD imposed its own groundwater allocations as early as 1978, and the Central Platte NRD imposed fertilizer use regulations that have successfully lowered nitrate contamination in that NRD's groundwater supplies.[46] In many cases, NRDs have been able to avoid overharvesting and groundwater contamination without having to implement regulations by educating and providing assistance to irrigators to reduce their use of groundwater and fertilizers. For example, because of such efforts by the Upper Big Blue NRD, groundwater levels are above what they were in 1961, despite the addition of more than 420,000 groundwater-irrigated acres.[47] Although there are areas of the state where groundwater levels are still declining and where nitrate levels are still high, because there is a time lag, often measured in years, between the initiation of a management action to protect groundwater and the impact of that action on groundwater, it is too soon to conclude whether existing NRD rules and regulations will be able to sustain Nebraska's groundwater supplies.[48]

In resilient institutions, sanctioning is not implemented by external authorities but by the participants themselves.[49] State law provides for graduated sanctions, and NRDs often allow irrigators a period of time to achieve compliance without imposing a penalty.[50]

According to Ostrom, rules specifying the quantity of a resource a user is allocated must be related to local conditions and to rules requiring labor, materials, or money inputs.[51] If the initial set of rules established by the users or by a government is not tailored to fit the local problem, or if the benefits derived from the resource do not outweigh the costs to use the resource, long-term sustainability may not be achieved.[52] With Nebraska's wide range of climate, topography, groundwater aquifers, and water uses, a single approach to water governance does not fit all situations. Because each NRD can make its own rules, each of the twenty-three NRDs has developed appropriation and provision rules to match its local physical and climatic conditions and water needs.[53] Moreover, because the NRDs are governed by locally elected boards that monitor their taxing authority, the system also provides a mechanism to ensure a proportional equivalence between the benefits and the costs of an NRD's operation.[54]

The long-term effectiveness of rules depends on monitoring harvesting practices and the ability of the users to understand and verify the results.[55] State law encourages, and in some instances requires, the NRD to monitor biophysical conditions and user behavior.[56] In large part because of their understanding of the importance of protecting the groundwater on which they personally depend, the NRDs have developed the largest database for monitoring nitrates and agricultural pesticides in groundwater in the United States.[57]

A stable funding source is also necessary for resilient governance.[58] When conditions are complex and unknown, as is often the case with groundwater,[59] and uncertainty is high, it is critical to have funding for the development of research that can accurately identify problems and design effective solutions.[60] Funding is also important to build infrastructure and provide incentives to alleviate these problems.[61] When created, the NRDs were given taxing authority. Additional legislation gave them authority to issue bonds and assess an irrigated acres tax.[62] The recently established Water Sustainability Fund provided a one-time startup fund of $21 million, and the dedication of $11 million per year, with no sunset clause.[63] The NRDs were an important political force that encouraged the legislature to provide a more stable and sufficient funding source to develop and sustain water management programs.[64]

Secure tenure arrangements empower people and provide the basis for investing in the future, which helps sustain the resource.[65] Insecure tenure arrangements, by contrast, work against effective water governance.[66] In Nebraska, state law provides certainty that surface water rights will not be harmed by other surface water appropriators, but it does not provide for secure tenure rights for surface water appropriators and groundwater-well owners adversely affected by the groundwater use. In such cases, the law relies on the implementation of an integrated management plan to protect these rights. Although many integrated management plans have been implemented, surface water appropriators still believe their rights have not been protected, and have resorted to litigation to protect their rights.[67]

Further exacerbating this problem, state law does not provide for rapid, low-cost, local access to conflict resolution mechanisms.[68] According to Ostrom, at the most local level, water users and their officials must have rapid access to low-cost, local arenas to resolve conflicts among users or between users and officials.[69] Without such mechanisms water users can be left feeling powerless and ineffective in their efforts to adequately and effectively manage the resource.[70] Although many disputes are resolved informally at the local level by both the state DNR and the NRDs, the remaining unresolved conflicts could perhaps be alleviated without the need for costly litigation by creating more local and lower-cost dispute resolution mechanisms, such as a process for nonbinding dispute arbitration.

A number of criteria for resilient governance, effective and efficient communication systems, leadership, trust, and collaborative decisionmaking can be enabled by a good legal framework, but laws alone cannot ensure these criteria will be met. Rather, meeting these criteria depends on the personal characteristics and actions of the people involved with the governance institution.

Of note, groups that do not communicate are more likely to overuse the resource.[71] Communication increases the potential for trust,[72] lowers the cost of monitoring behavior, and induces rule compliance.[73] Simply allowing "cheap" talk enables people to reduce overharvesting and increase joint payoffs.[74]

Good leadership, which involves making difficult choices that are in the best interest of society, providing overarching direction to constituents, and being willing to be a part of the long-term decisionmaking process, is essential to shaping change, managing conflict, linking actors, initiating partnerships among actor groups, compiling and generating knowledge, and mobilizing broad support for change.[75] Trust is the basis of all social institutions.[76] Only when there is trust can governance institutions work well over time.[77] With trust comes reciprocity and cooperation, which lower the transaction costs in reaching agreements and induce rule compliance, which in turn lowers the costs of monitoring and enforcement.[78] When the parties do not have trust among themselves, fragmentation and conflicts are more likely.[79]

These attributes in turn encourage collective-choice arrangements in which most individuals affected by harvesting and protection rules are included in the group that can modify these rules. According to Ostrom, governance institutions that collaborate well are better able to tailor their rules to local circumstances because the individuals who directly interact with one another and the physical world are in the best position to fit the rules to their specific setting.[80] Collaboration also increases knowledge, results in more creative solutions, and increases trust; and once appropriators have committed to the rules, they are motivated to monitor and help ensure the compliance of other appropriators. Rules established by the resource user are also better understood, and are more likely to be perceived as being legitimate, which helps prevent legal challenges during later stages of the decision process.[81]

Collaboration also increases fairness and equity.[82] Despite differences in how people use and value water, it is essential that all water users feel they are treated fairly.[83] Generally, two forms of equity have been emphasized in the literature: distribution justice, which emphasizes a fair distribution of impacts, benefits, and costs, and participatory justice, which stresses procedures that provide for the fair involvement of all parties in decisionmaking.[84] Ensuring that procedural decisions are made on a level playing field where both institutional and stakeholders' concerns are taken into account reduces the chances that the resource users will try to challenge, avoid, or disrupt the policies of the governing institution.

NRDs have different boards, and different leaders. Hence the extent to which an individual NRD exhibits the above characteristics varies among NRDs. Where NRD rules have been the result of a collaborative and transparent process, most water users trust the NRD and believe they have been fairly treated. Where such processes were not the norm, water users, particularly surface water users and environmental groups, do not trust the NRD or the state DNR and do not believe they were treated fairly.[85]

Adaptive management, often characterized as "learning by doing," is a formal iterative management process that requires defining the problem, identifying clear objectives, formulating evaluation criteria, estimating outcomes, evaluating trade-offs, deciding on a plan of action, implementing the plan, monitoring the results, evaluating the success of the actions, and adjusting the plan as necessary to achieve the results.[86] Adaptive management was developed to cope with the surprises and uncertainties of ecosystem changes. It is particularly useful when there is uncertainty resulting from environmental variation, difficulty in observing the status of the resource, incomplete controllability, and a lack of understanding of the underlying system processes.[87] In an adaptive management process, policies are treated as hypotheses and all management can be seen as a kind of hypothesis testing.[88] Unexpected outcomes are seen not as failures but as an opportunity for learning.[89] The involvement of representative stakeholders in all steps of the process is also a key component of adaptive management.[90] Although Nebraska law does not explicitly use the term "adaptive management," the law requires all the basic steps of the adaptive management process when implementing an integrated management plan.[91]

In the past, simple strategies for governing the world's resources that relied exclusively on one-level centralized command and control often failed, sometimes catastrophically.[92] In today's more complex society, governance activities are best organized in nested enterprises in which appropriation, monitoring, enforcement, conflict resolution, and other governance activities are organized in multiple layers.[93] Furthermore, a nested enterprise can ensure that the allocation and management of water resources across upstream and downstream regions do not create harmful impacts to others without mitigation or compensation.[94] Ostrom found that establishing rules at one level without rules at the other levels will produce an incomplete system that may not endure over the long term.[95]

The locally controlled NRDs are clearly part of a nested hierarchy in which, at least to some extent, the NRDs are controlled at the state level.[96] For the most part, this system has worked well. However, in the interest of maintaining local control, the governance of some surface water river basins and groundwater aquifers was split among more than one NRD. Thus the impacts of the governance of one NRD can affect the water users in another NRD. This problem was recognized as an issue in Nebraska's law with intent language stating that "the Legislature intends and expects that each natural resources district will accept responsibility for groundwater management . . . in the same manner and to the same extent as if the impacts were contained within that district."[97] However, as recently pointed out by both the Nebraska Supreme Court[98] and the U.S. Supreme Court,[99] the law did not provide the state with any enforceable mechanisms to ensure that the intent of the law was followed. Hence the hierarchical system is incomplete. Nevertheless, the nested hierarchy in Nebraska has provided many of the advantages of local control mentioned above.

Polycentric governance provides for multiple centers of governance with overlapping power[100] and allows local associations to work together through bridging organizations with larger governmental providers of infrastructure and resources.[101] Polycentric institutions cross administrative boundaries; create opportunities for understanding and for servicing needs in spatially heterogeneous contexts; can be important for handling scale-dependent interactions; allow economies of scale in dividing tasks across government bodies; and, through collaboration and cooperation among various governance institutions, can increase citizen involvement, learning, and levels of trust between organizations. All of these affordances lower transaction costs and create a venue for resolving conflicts, enabling legislative polices, and increasing creativity.[102]

Adaptive comanagement combines the emphasis on learning and experimentation of adaptive management with the emphasis on comanagement, or the sharing of rights, responsibilities, and power between the different levels, found in polycentric governance systems.[103] Because the redundancy inherent in polycentric governance limits the risk of experimentation, an adaptive comanagement system can afford to treat policies as hypotheses and management actions as experiments. Adaptive comanagement also implies a focus on the bioregion, which, in water management, often translates into management at the basin level.[104]

In Nebraska, the local NRDs are part of a nested hierarchical, polycentric governance system that, in addition to state government, includes cities, towns, counties, and at times the federal government. There is a high degree of overlap of problems and a mixture of authorities among these units. Through interlocal agreements, which were authorized by the Interlocal Cooperation Act,[105] the NRDs are able to establish formal, legally described, bridging organizations with cities, counties, and others to work on issues of mutual concern. For example, the Lower Platte South NRD, together with the University of Nebraska and the City of Lincoln, created an organization to collaborate on building a major road, constructing a flood-control project, and developing a recreational trail system through the heart of the city.[106] The NRDs along the Platte River undertook the Cooperative Hydrology Study (COHYST) project[107] to develop water models to manage and regulate Platte River water use. The Lower Platte South NRD, along with two other NRDs, six state agencies, and many cities and towns, formed the Lower Platte River Corridor Alliance, which is implementing adaptive management to deal with a wide range of water and natural resource–related issues in the Lower Platte River basin, one of the most heavily populated and fastest-growing areas of Nebraska.[108] Also, many NRDs are working with cities and towns to protect the quality of their groundwater supplies or to develop water-supply projects.[109] The use of such bridging organizations has encouraged participation by a wide range of public interest groups and governing entities and has made it possible to tailor governance to the specific problem at the proper scale. Because participation is voluntary, these bridging organizations are respected by the individual governance entities and do not suffer from the issues of legitimacy and lack of collaborative decisionmaking that often seem to plague more top-down river basin governance authorities.[110]

Assessment of Nebraska's NRD Governance System's Ability to Adapt to Climate Change

Not surprisingly, many of the characteristics that have defined resilient water governance systems in the past are the ones likely to be important for providing our legal and governance institutions with the capacity needed to adapt to climate change in the future. The NRD governance system already meets many of these criteria.

Arnold and Gunderson, Cosens and Stow, and Green and Perrings all argue that governance institutions must transcend artificial and political boundaries and holistically address interrelated water issues at watershed scales.[111] The NRD boundaries do transcend artificial boundaries, and NRDs can holistically address a wide range of water issues.

Cosens and Stow also state that the capacity of local governance entities must be increased.[112] Esty observed that local governance increases accountability, and that the accountability gap increases with increases in scale of government.[113] The NRDs are clearly enabled by state law to implement local governance, and because they are governed by locally elected boards, they have local accountability to the district's electorate.

Cosens and Stow also state that local entities must be provided with substantial resources, and that funding should be greater at the bottom than the top of the governance hierarchy.[114] The NRDs have local taxing authority, and with the recent enactment of the Water Sustainability Fund, there should be substantial funds available to local entities.

According to Camacho, uncertainty is the greatest challenge.[115] Therefore, to achieve resilience, it is necessary to recognize the limits of available knowledge and create space for both technical and policy experimentation and innovation. The adaptive management planning process was developed specifically to increase the resilience of socioecological systems to the stresses, instability, and uncertainty that are more characteristic of these systems than stability.[116] Climate change makes the use of adaptive management even more important.[117] Under Nebraska's legal framework, basins that are determined to be legally fully or overly appropriated are required to develop an integrated management plan, which in turn requires many of the steps that are part of the adaptive management process. Moreover, the fact that there are twenty-three NRDs, each developing its own plans, means there is a lot of opportunity for the different NRDs to experiment, and to learn from one another's experiments.

Camacho, Cosens and Stow, and Karkkainen all argue that the key to success is to monitor the results of regulatory strategies and adjust the regulations as necessary to achieve the goals of the plan.[118] The NRDs already make extensive use of water-monitoring programs and use the results of this monitoring to trigger actions to address water problems and to test the efficacy of these actions.[119]

Because collective choice and collaboration are particularly important when dealing with uncertainty, Cosens and Stow emphasize the salience of communication, leadership, trust, and collaboration to allow governance institutions to cope with climate change.[120] According to Hamm and his colleagues, attention to fair procedures bears special weight when decisions must be made in the face of uncertainty.[121] As mentioned earlier, the success of these criteria depends primarily on the personalities and character traits of the people involved in government, and therefore success varies among the NRDs.

Ostrom advocates polycentric responses to large-scale complex environmental problems such as climate change.[122] Arnold and Gunderson also emphasize the importance of polycentric governance. They argue that monocentric governance that uses rules that are resistant to change will be maladaptive in the face of climate change. By contrast, polycentric governance utilizes multimodal and multiscalar responses to problems, embraces a multiplicity and diffusion of governmental authority, creates space for experimentation and innovation; enables risk and diversification (a key element of resilience), and provides redundancy to absorb shocks, all of which will enable adaptive capacity to climate change.[123] Arnold and Gunderson point out, however, that the legal system must provide both flexibility and stability, as well as appropriate and relevant overarching standards, to govern the exercise of discretion at the local level and to ensure local decisionmakers are accountable.[124] In Nebraska the NRDs are embarking on polycentric governance and, through legally drafted interlocal agreements, are forming bridging organizations with cities, counties, nonprofit organizations, and state and federal agencies to address problems in their district and in the basin as a whole.[125] Furthermore, adding to the redundancy inherent in polycentric governance systems, the twenty-three NRDs, each developing its own rules, provide more redundancy in the system.

In sum, to cope with climate change, there are advantages to relying on decentralized overlapping authority, which leads to a diversity of focused localized strategies that promote regulatory experimentation and opportunities for jurisdictional learning.[126] The NRD legal framework provides and encourages such a system. Moreover, the locally controlled NRDs increasingly are implementing adaptive management and, perhaps more important, adaptive comanagement. These strategies, which have already enabled the NRDs to cope with today's water problems, will no doubt help give them the flexibility and adaptive capacity needed to cope with future challenges and the uncertainties of climate change. To the extent the NRDs succeed in these efforts, Nebraska's cities and towns will have the benefit of a sufficient supply of good-quality water.

Notes

1. Ann Bleed and Christina Hoffman Babbitt, *Nebraska's Natural Resources Districts: An Assessment of a Large-Scale Locally Controlled Water Governance Framework*, Policy Report of the Robert B. Daugherty Water for Food Institute (University of Nebraska, 2015).

2. U.S. Geological Survey. *Estimated Use of Water in the United States in 2010. Water Use in Nebraska: Total Fresh Groundwater Use, 2010* (Department of the Interior, 2015) (http://pubs.usgs.gov/circ/1405/pdf/circ1405.pdf).

3. Noel Gollehon and Bernadette Winston, *Groundwater Irrigation and Water Withdrawals: The Ogallala Aquifer Initiative*, U.S. Department of Agriculture, Natural Resources Conservation Service, Economic Series 1 (Department of the Interior, August 2013) (https://prod.nrcs.usda.gov/Internet/FSE_DOCUMENTS /stelprdb1186440.pdf).

4. Nebraska Department of Health and Human Services, Division of Public Health, *Nebraska's Public Water System Program 2013 Annual Report* (2014) (http:// dhhs.ne.gov/publichealth/Documents/PublicWaterSupplyAnnualReport.pdf).

5. J. B. Ruhl, "Climate Change Adaptation and the Structural Transformation of Environmental Law," *Environmental Law* 40 (2010), pp. 363–435; Tarsha Eason, Alyson C. Flournoy, Heriberto Cabezas, and Michael A. Gonzalez, "A Case for Preserving a Natural Resource Legacy and Promoting a Sustainable Future," in *Social-Ecological Resilience and Law*, edited by Ahjond S. Garmestani and Craig R. Allen (Columbia University Press, 2014), pp. 293–316.

6. Holly Doremus and Michael Hanemann, "The Challenges of Dynamic Water Management in the American West," *Journal of Environmental Law and Policy* 25, no. 2 (2008), pp. 54–75; J. B. Ruhl, "Climate Change Adaptation"; Ahjond S. Garmestani, Craig R. Allen, J. B. Ruhl, and C. S. Holling, "The Integration of Social-Ecological Resilience and Law," in *Social-Ecological Resilience and Law*, edited by Garmestani and Allen, pp. 365–82.

7. Ruhl, "Climate Change Adaption"; Eason and others, "A Case for Preserving a Natural Resouce Legacy."

8. Ruhl, "Climate Change Adaptation"; Arnold, Craig Anthony, and Lance H. Gunderson, "Adaptive Law," in *Social-Ecological Resilience and Law*, edited by Garmestani and Allen, pp. 317–64.

9. Brian Walker and David Salt, *Resilience Thinking: Sustaining Ecosystems and People in a Changing World* (Washington, D.C.: Island Press, 2006), p. xiii.

10. A socioecological system refers to the fact that we all live in social-economic systems that are inextricably linked with ecological systems (Walker and Salt, *Resilience Thinking*).

11. Walker and Salt, *Resilience Thinking*; Ruhl, "Climate Change Adaptation"; Barbara Cosens and Craig A. Stow, "Addressing Fragmentation and Uncertainty in Water Allocation and Water Quality Law," in *Social-Ecological Resilience and Law*, edited by Garmestani and Allen, pp. 142–75.

12. Elinor Ostrom, *Governing the Commons: The Evolution of Institutions for Collective Action* (Cambridge University Press, 1990); idem, *Understanding Institutional Diversity* (Princeton University Press, 2005); idem, "Prize Lecture by Elinor Ostrom," Aula Magna, Stockholm University, December 8, 2009, video, NobelPrize.org. (www.nobelprize.org/mediaplayer/index.php?id=1223).

13. Ostrom, *Governing the Commons*.

14. J. M. Anderies, M. A. Janssen, and E. Ostrom, "A Framework to Analyze the Robustness of Social-ecological Systems from an Institutional Perspective," *Ecology and Society* 9, no. 1 (2004), art. 18 (www.ecologyandsociety.org/vol9/iss1/art18).

15. Hazel Jenkins, "A History of Nebraska's Natural Resources Districts," edited by Robert B. Hyer, 1975, unpublished manuscript.

16. Neb. Rev. Stat. § 2-3229 (2016).

17. Neb. Rev. Stat. § 2-3229 (2016).

18. Nebraska Legislature 1975 L.B. 577. Laws of Nebraska 84th Legislature First Session 1975, Nebraska Secretary of State Office, pp. 1145–58.

19. Stephen D. Mossman, "'Whiskey Is for Drinkin' but Water Is for Fightin' About': A First-hand Account of Nebraska's Integrated Management of Ground and Surface Water Debate and the Passage of L.B. 108," *Creighton Law Review* 30 (1996), pp. 67–104.

20. Nebraska Department of Natural Resources, *Report of the Nebraska Water Policy Task Force to the 2003 Nebraska Legislature* (Lincoln, 2003) (http://www.dnr .nebraska.gov/Media/PDF/TaskForceFinalReport1-02-04.pdf).

21. Neb. Rev. Stat. § 46, art. 7 (2016).

22. Nebraska Department of Natural Resources, *Report of the Nebraska Water Policy Task Force*.

23. Neb. Rev. Stat. § 46-715(2)(a) (2016).

24. Neb. Rev. Stat. § 46-715(3) (2016).

25. Neb. Rev. Stat. § 46-715(4) (2016).

26. Jennifer Schellpeper, Integrated Water Management Division Head, e-mail August 15, 2016.

27. Bleed and Babbitt, *Nebraska's Natural Resource Districts*.

28. The High Plains Aquifer is a group of hydrologically connected aquifers, including the Ogallala Aquifer, and both older and younger geological formations (V. L. Maguire, M. R. Johnson, R. L. Schieffer, J. S. Stanton, S. K. Sebree, and I. M. Verstraeten, "Water in Storage and Approaches to Groundwater Management, High Plains Aquifer, 2000," U.S. Geological Survey Circular 1243 [2003]).

29. Jesse T. Korus, Leslie M. Howard, Aaron R. Young, Dana P. Divine, Mark E. Burbach, J. Michael Jess, and Douglas R. Hallum, *The Groundwater Atlas of Nebraska*, Resource Atlas No. 4b/2013, 3rd (rev.) ed. (University of Nebraska–Lincoln, School of Natural Resources, 2013).

30. Maguire and others, *Water in Storage and Approaches to Groundwater Management*.

31. Korus and others, *The Groundwater Atlas of Nebraska*; see also University of Nebraska Conservation and Survey Division map, "Groundwater-Level Changes in Nebraska—Predevelopment to Spring 2015" (http://snr.unl.edu/data/water /groundwater/gwlevelchangemaps.aspx).

32. U.S. Environmental Protection Agency, *An Evaluation of the Clean Water Act Section 319 Program* (2011) (http://water.epa.gov/polwaste/nps/upload/319evaluation .pdf).

33. Lincoln Water System, Lincoln Water System Draft Facilities Master Plan Stakeholder Committee, City of Lincoln Project 701353 (2013).

34. Deborah J. Bathke, Robert J. Oglesby, Clinton M. Rowe, and Donald A. Wilhite, *Understanding and Assessing Climate Change: Implications for Nebraska. A Synthesis Report to Support Decision Making and Natural Resource Management in a Changing Climate* (University of Nebraska–Lincoln, 2014).

35. Bleed and Babbitt, *Nebraska's Natural Resources Districts*.

36. John Wesley Powell, "Institution for the Arid Lands," *Century* 18 (1890), cited in Walter Prescott Webb, *The Great Plains* (New York: Ginn and Co., 1931); J. B. Ruhl, C. L. Lant, S. E. Kraft, J. Adams, L. Duram, and T. Loftus, "Proposal for a Model State Watershed Management Act," *Environmental Law (Lewis and Clark Law School)* 4, no. 37 (2003).

37. M. S. Rosegrant, X. Cai, and S. A. Cline, *Global Water Outlook to 2025: Averting an Impending Crisis* (Washington, D.C.: International Food Policy Research Institute/International Water Management Institute, 2002).

38. Bleed and Babbitt, *Nebraska's Natural Resources Districts*.

39. Ibid.

40. Ostrom, *Governing the Commons*; idem, "A General Framework for Analyzing Sustainability of Social-Ecological Systems," *Science* 325, no. 5939 (2009), pp. 419–22.

41. Ostrom, *Governing the Commons.*

42. Anderies and others, "Framework"; Ostrom, *Governing the Commons;* idem, "A General Framework."

43. Ostrom, *Governing the Commons.*

44. Bleed and Babbitt, *Nebraska's Natural Resources Districts.*

45. Ibid.

46. Ibid.

47. Ibid.

48. Ibid.

49. Ostrom, *Governing the Commons;* idem, "Prize Lecture."

50. Bleed and Babbitt, *Nebraska's Natural Resources Districts.*

51. Ostrom, *Governing the Commons.*

52. Ibid.; Ostrom, "A General Framework"; idem, "Prize Lecture"; Anderies and others, "Framework."

53. Bleed and Babbitt, *Nebraska's Natural Resources Districts.*

54. Ibid.

55. Thomas Dietz, Elinor Ostrom, and Paul C. Stern, "The Struggle to Govern the Commons," *Science* 302, no. 5652 (2003), pp. 1907–12; Ostrom, "A General Framework."

56. Bleed and Babbitt, *Nebraska's Natural Resources Districts.*

57. Marty Link, water resource specialist, State Department of Environmental Quality, personal interview, July 2014.

58. Carl Folke, Thomas Hahn, Per Olsson, and Jon Norberg, "Adaptive Governance of Social-ecological Systems," *Annual Review of Environmental Resources* 30 (2005), pp. 441–73; Holly Doremus, Alejandro E. Camacho, Daniel A. Farber, Robert L. Glicksman, Dale D. Goble, William L. Andreen, Dan Rohlf, Bradley C. Karkkainen, Shana Campbell Jones, A. Dan Tarlock, Sandra B. Zellmer, and Ling-Yee Huang, "Making Good Use of Adaptive Management," Center for Progressive Reform White Paper 1104, UC Irvine School of Law Research Paper 2011–24 (2011) (http://ssrn.com/abstract=1808106 or http://dx.doi.org/10.2139/ssrn.1808106); Christina Hoffman and Sandra Zellmer, "Integration and Adaptability: Assessing Institutional Ability for Adaptive Water Resources Management," *Nebraska Law Review* 91 (2013), pp. 805–65.

59. Bleed and Babbitt, *Nebraska's Natural Resources Districts.*

60. Dietz and others, "The Struggle to Govern the Commons."

61. Ibid.

62. Neb. Rev. Stat. § 2-3226 (2016).

63. Neb. Rev. Stat. § 61-222 (2016).

64. Karen Griffin, groundwater technical leader, Olsson Associates, personal interview, May, 2014. Karen Griffin was the project manager for the Water Funding Task Force.

65. Ostrom, *Governing the Commons*; International Bank for Reconstruction and Development/World Bank, *Reengaging in Agricultural Water Management Challenges and Options* (Washington, D.C.: World Bank, 2006); Ruth Meinzen-Dick, "Beyond Panaceas in Water Institutions," *Proceedings of the National Academy of Sciences* 104, no. 39 (2007), pp. 15200–205; Chris Perry, "Water Governance and the Role of Tenure and Rights in Coping with Agricultural Water Scarcity," Background Paper One, Expert Consultation, Food and Agriculture Organization of the United Nations, Rome, Italy, January 21–23, 2013.

66. Stephen Hodgson, "Water Governance and the Role of Tenure and Rights in Coping with Agriculture Water Scarcity," Background Paper Three, Expert Consultation, Food and Agriculture Organization of the United Nations, Rome, Italy, January 20–22, 2013.

67. Spear T Ranch, Inc. v. Knaub, 691 N.W. 2d 116 (2005), 269 Neb. 177; Hill, Coffey, Uerling, Schaffert v. State of Nebraska and Nebraska Department of Natural Resources (2014), Domina Law Firm (http://www.dominalaw.com/documents/Irrigators-v.-State-of-Nebraska.pdf); "Irrigators File New Lawsuit," *Hastings Tribune*, November 4, 2015 (www.hastingstribune.com/news/bostwick-irrigation-district-sues-state-nrds/article_b917583a-bfa0-11e5-93e6-b3e79c157143.html); "Bostwick Irrigation District Sues State, NRDs," *Hastings Tribune*, January 20, 2016 (http://www.mccookgazette.com/story/2247323.html).

68. Bleed and Babbitt, *Nebraska's Natural Resources Districts*.

69. Ostrom, *Governing the Commons*.

70. Ibid.; Ostrom, "A General Framework."

71. Ostrom, *Governing the Commons*.

72. Folke and others, "Adaptive Governance of Social-ecological Systems."

73. Dietz and others, "The Struggle to Govern the Commons."

74. Elinor Ostrom, "Beyond Markets and States: Polycentric Governance of Complex Economic Systems." *American Economic Review* 100, no. 3 (2010), pp. 641–72.

75. Folke and others, "Adaptive Governance of Social-ecological Systems"; Ostrom, "A General Framework."

76. Folke and others, "Adaptive Governance of Social-ecological Systems."

77. Ostrom, "Prize Lecture"; idem, "Reflections on 'Some Unsettled Problems of Irrigation,'" *American Economic Review* 101 (February 2011), pp. 49–63 (www.aeaweb.org/articles.php?doi=10.1257/aer.101.1.49).

78. Ostrom, *Governing the Commons*; Dietz and others, "The Struggle to Govern the Commons"; Anderies and others, "Framework"; J. M. Anderies, B. H. Walker, and A. P. Kinzig, "Fifteen Weddings and a Funeral: Case Studies and Resilience-Based Management," *Ecology and Society* 11, no. 1 (2006), art. 21 (http://www.ecologyandsociety.org/vol11/iss1/art21/); J. A. Hamm, L. M. PytlikZillig, M. N. Herian, A. J. Tomkins, H. Dietrich, and S. Michaels, "Trust and Intention to Comply with a Water Allocation Decision: The Moderating Roles of Knowledge and Consistency,"

Ecology and Society 18, no. 4 (2013), art. 49 (http://dx.doi.org/10.5751/ES-05849 -180449).

79. D. Huitema, E. Mostert, W. Egas, S. Moellenkamp, C. Pahl-Wostl, and R. Yalcin, "Adaptive Water Governance: Assessing the Institutional Prescriptions of Adaptive Co-management from a Governance Perspective and Defining a Research Agenda," *Ecology and Society* 14, no. 1 (2009), art. 26 (http://www.ecologyandsociety .org/vol14/iss1/art26/).

80. Ostrom, *Governing the Commons.*

81. Ibid.; Anderies and others, "Framework"; Dietz and others, "The Struggle to Govern the Commons"; L. Lebel, J. M. Anderies, B. Campbell, C. Folke, S. Hatfield-Dodds, T. P. Hughes, and J. Wilson, "Governance and the Capacity to Manage Resilience in Regional Social-ecological Systems," *Ecology and Society* 11, no. 1 (2006), art. 19 (www.ecologyandsociety.org/vol11/iss1/art19/); Firket Birkes, "Evolution of Co-management: Role of Knowledge Creation, Bridging Organizations, and Social Learning," *Journal of Environmental Management* 90 (2009), pp. 1692–702; Huitema and others, "Adaptive Water Governance."

82. S. M. Langsdale, A. Beall, J. Carmichael, S. J. Cohen, C. B. Forster, and T. Neale, "Exploring the Implications of Climate Change on Water Resources through Participatory Modeling: Case Study of the Okanagan Basin, British Columbia," *Journal of Water Resources Planning and Management* 135, no. 5 (2009), pp. 373–81; P. Reed and J. Kasprzyk, "Water Resources Management: The Myth, the Wicked, and the Future," *Journal of Water Resources Planning and Management* 135 (2009), pp. 411–13; Arnim Wiek and Kelli L. Larson, "Water, People, and Sustainability: A Systems Framework for Analyzing and Assessing Water Governance Regimes," *Water Resources Management* 26 (2012), pp. 3153–71 (doi 10.1007/s11269-012-0065-6).

83. Ostrom, *Governing the Commons;* idem, " General Framework"; G. Syme, B. Nancarrow, and J. McCreddin, "Defining the Components of Fairness in the Allocation of Water to Environmental and Human Uses," *Journal of Environmental Management* 57, no. 1 (1999), pp. 51–70.

84. Wiek and Larson, "Water, People, and Sustainability"; J. M., Anderies, M. A. Janssen, and E. Ostrom. "A Framework to Analyze the Robustness of Social-Ecological Systems from an Institutional Perspective," *Ecology and Society* 9, no. 1 (2004), p. 18 (http://www.ecologyandsociety.org/vol9/iss1/art18); Joseph Hamm, "Understanding the Role of Trust in Cooperation with Natural Resources Institutions." Unpublished doctoral dissertation. University of Nebraska–Lincoln, Nebraska (http://digitalcommons.unl.edu/psychdiss/63/).

85. Bleed and Babbitt, *Nebraska's Natural Resource Districts;* C. Hoffman Babbitt, M. Burbach, and L. Pennisi, "A Mixed Methods Approach to Assessing Success in Transitioning Water Management Institutions: A Case Study of the Platte

River Basin, Nebraska," *Ecology and Society* 20, no. 1 (2015), art. 54 (www.ecology andsociety.org/vol20/iss1/art54/).

86. C. R. Allen, T. J. Fontaine, K. L. Pope, and A. S. Garmestani, "Adaptive Management for a Turbulent Future," *Journal of Environmental Management* 92 (2011), pp. 1339–45.

87. Ibid.

88. Huitema and others, "Adaptive Water Governance."

89. Ibid.

90. Folke and others, "Adaptive Governance of Social-ecological Systems."

91. Neb. Rev. Stat. § 46-715(2) and § 46-715(4).

92. M. Hajer, "Policy without Polity: Policy Analysis and the Institutional Void," *Policy Sciences* 36 (2003), pp. 175–95.

93. Lebel and others, "Governance and the Capacity to Manage Resilience"; Ostrom, "Prize Lecture."

94. Wiek and Larson, "Water, People, and Sustainability."

95. Ostrom, *Governing the Commons.*

96. Bleed and Babbitt, *Nebraska's Natural Resources Districts.*

97. Neb. Rev. Stat. § 46-703(4).

98. Spear T Ranch, Inc. v. Knaub (2005).

99. U.S. Supreme Court, Kansas v. Nebraska and Colorado (2015); U.S. Supreme Court, No. 126 Orig. State of Kansas, Plaintiff, v. States of Nebraska and Colorado on Exceptions to Report of Special Master (February 24, 2015).

100. Huitema and others, "Adaptive Water Governance."

101. Meinzen-Dick, "Beyond Panaceas in Water Institutions."

102. Dietz and others, "The Struggle to Govern the Commons"; Ostrom, *Understanding Institutional Diversity;* Lebel and others, "Governance and the Capacity to Manage Resilience"; Huitema and others, "Adaptive Water Governance."

103. Berkes, "Evolution of Co-management"; Huitema and others, "Adaptive Water Governance"; Allen and others, "Adaptive Management for a Turbulent Future."

104. Huitema and others, "Adaptive Water Governance."

105. Neb. Rev. Stat. § 13-804.

106. Bleed and Babbitt, *Nebraska's Natural Resources Districts.*

107. COHYST, Platte River Cooperative Hydrology Study (COHYST), Nebraska Department of Natural Resources (http://cohyst.dnr.ne.gov/).

108. Lower Platte Corridor Alliance, website (2014) (http://www.lowerplatte.org/); Meghan Sittler, coordinator, Lower Platte River Corridor Alliance, personal interview, March 2014.

109. Bleed and Babbitt, *Nebraska's Natural Resource Districts.*

110. Christopher Lant, "Watershed Governance in the United States: The Challenges Ahead," Universities Council on Water Resources, *Water Resources Update*

126 (2003), pp. 21–28 (http://ucowr.org/files/Achieved_Journal_Issues/126 _A3Watershed%20Governance%20in%20the%20United%20States.pdf).

111. Arnold and Gunderson, "Adaptive Law"; Cosens and Stow, "Addressing Fragmentation and Uncertainty in Water Allocation and Water Quality Law"; Olivia Odom Green and Charles Perrings, "Institutionalized Cooperation and Resilience in Transboundary Freshwater Allocation," in *Social-Ecological Resilience and Law*, edited by Garmestani and Allen, pp. 176–203.

112. Cosens and Stow, "Addressing Fragmentation and Uncertainty in Water Allocation and Water Quality Law."

113. D. C. Esty, "Good Governance at the Supranational Scale: Globalizing Administrative Law," *Yale Law Journal* 115 (2006), pp. 1490–1562.

114. Cosens and Stow, "Addressing Fragmentation and Uncertainty in Water Allocation and Water Quality Law."

115. A. E. Camacho, "Adapting Governance to Climate Change: Managing Uncertainty through a Learning Infrastructure," *Emory Law Journal* 59 (2009), pp. 1–77.

116. Walker and Salt, *Resilience Thinking*.

117. Folk and others, "Adaptive Governance of Social-ecological Systems"; Huitema and others, "Adaptive Water Governance."

118. Camacho, "Adapting Governance to Climate Change"; Cosens and Stow, "Addressing Fragmentation and Uncertainty in Water Allocation and Water Quality Law"; B. C. Karkkainen, "Bottlenecks and Baselines: Tackling Information Deficits in Environmental Regulation," *Texas Law Review* 86 (2008), pp. 1409–44.

119. Bleed and Babbitt, *Nebraska's Natural Resources Districts*.

120. Cosens and Stow, "Addressing Fragmentation and Uncertainty in Water Allocation and Water Quality Law."

121. Hamm and others, "Trust and Intention to Comply with a Water Allocation Decision."

122. Elinor Ostrom, "A Polycentric Approach for Coping with Climate Change," World Bank Research Working Paper 5095 (Washington, D.C.: World Bank, 2009).

123. Arnold and Gunderson, "Adaptive Law."

124. Ibid.

125. Bleed and Babbitt, *Nebraska's Natural Resources Districts*.

126. A. E. Camacho, "Transforming the Means and Ends of Natural Resources Management," *North Carolina Law Review* 89 (2011), pp. 1405–50.

Chapter 6

Groundwater in the American West
How to Harness Hydrogeological Analysis to Improve
Groundwater Management

BURKE W. GRIGGS AND JAMES J. BUTLER JR.

The West is running out of groundwater. Across the High Plains–
Ogallala Aquifer, the largest source of fresh water in the West, irrigators
are pumping as much as 19 million acre-feet per year.[1] Since the onset of
groundwater pumping for irrigation, the aquifer has lost over 276 million
acre-feet—enough to put all of Kansas under five feet of water.[2] Across the
Colorado River basin, the main source of water for 40 million westerners,
groundwater supplies connected to the river plummeted by 41 million
acre-feet from 2004 to 2013.[3] That is enough water to satisfy the domestic needs of the entire U.S. population for eight years.[4] And in the
Central Valley of California, the most valuable agricultural area in the
United States, intense groundwater pumping has drained 125 million acre-
feet from local aquifers—enough water to fill the nation's largest reservoir, Lake Mead, nearly five times—while dropping land levels by as
much as a foot per year, and permanently reducing the aquifer's storage
space.[5] Most of this water is not coming back.

The causes of this crisis are not mysterious. The first has to do with
property rights: westerners still have the legal right to pump more groundwater than the aquifers can sustainably provide. In states that require a

distinct groundwater right or a permit, such as Colorado, Kansas, New Mexico, and Wyoming, such a right or permit constitutes a separate and permanent real property right.[6] Most rights to the Ogallala were granted between the early 1950s and the late 1970s, when there appeared to be enough groundwater for everyone. But the permanent decline of western aquifers now threatens the permanence of the water rights that depend on them. In other states, such as California, Nebraska, Oklahoma, and Texas, the right to use groundwater generally comes with the ownership of the overlying land. Landowners in these states are entitled to pump reasonable amounts of groundwater, but an irrigator's conception of what is reasonable rarely accords with a hydrologist's. As a consequence, there are far more water rights, permits, and pumping—all perfectly legal—than there is water to supply them, a condition known as overappropriation.

Generally, the owners of these rights have been reluctant to protect themselves against each other. Such reluctance seems odd, but it has its reasons, which lie behind the second cause of the groundwater crisis: the ways in which groundwater rights are—or are not—regulated. State laws rarely impose a duty to limit groundwater pumping to sustainable rates. State regulators and local groundwater districts are wary of reducing groundwater rights because of economic and political pressures to maintain irrigation at present levels, and seek to avoid lawsuits alleging that such reductions would amount to uncompensated (and therefore unconstitutional) takings of private property rights. And where owners do take legal action to reduce the pumping of neighbors whose use is impairing their own, the consequences can be uncertain, draconian, and expensive, not to mention unneighborly.

Yet the need to reduce groundwater pumping has become imperative, because the permanent depletion of western aquifers is threatening the West's future. Global climate change is disrupting precipitation patterns, reducing the snowpack on which cities, farms, and ecosystems depend, and threatening the desertification of millions of acres of fertile farm land.[7] Extended droughts across the West have reduced reservoir levels to record lows, forcing both cities and states to spend billions on water pipelines and millions on water lawsuits. These shortages are placing a justified premium on groundwater supplies—which are protected from drought and evaporation and stored at no cost underground, unlike surface reservoirs—just as those supplies are dropping at a distressingly rapid rate. Unrelenting migration to western cities and suburbs is forcing large-scale transfers of irrigation water to supply those population centers with water and power, drying up farms and impoverishing rural communities in the process. Even the federal government has become concerned about the wide-ranging effects of groundwater depletion, especially on the West's national forests, which hold rights to necessary water supplies under federal law.[8] Western states have responded with orchestrated horror to the specter of federal intrusion into their state law water domains, forcing the federal government to retreat.[9]

Policymakers recognize this imperative need. They comprehend the hydrological reality of permanent groundwater depletion, but their efforts at reform repeatedly run into the same man-made obstacles, which appear to be just as daunting and undeniable. There are the economic obstacles to reform, especially in agriculture, which uses between 80 and 90 percent of the West's water supplies but depends on cheap water. Local irrigation interests—irrigators, banks, and equipment and input suppliers—generally favor present rates of overpumping to maximize present economic yields. By their reckoning, the net present value of money exceeds the future value of groundwater, and so the long-term conservation of groundwater is economically irrational and even wasteful. Then there are the legal obstacles to reform: the property right in western water, which can be as peculiar as it is constitutionally protected, and the holders of groundwater rights, who will fight to maintain them at their face value, especially against regulatory reductions. These holders typically bridle at the notion of groundwater as a public resource because they perceive the water beneath their ground as their own. That perception has a powerful political reality, regardless of what the law says.[10]

Caught between hydrological facts on one side and the operative obstacles of economics, law, and politics on the other, water policy leaders have typically employed two coping mechanisms. Scientists and engineers tend to pretend that these obstacles do not exist, and so they plow ahead with purely technical solutions. Economists, bureaucrats, and lawyers tend to be cowed by these obstacles, and so they issue gradualist recommendations. We know that the problem of groundwater depletion is serious, and we are probably past the time for gradualist solutions. Together, experts have devised many policy answers to address the groundwater crisis. These include the protection of groundwater supplies, regulatory and voluntary reductions in groundwater pumping, hard floors on depletion levels, aquifer recharge during wet years and high river flows, better market mechanisms for transferring water rights between users and for rewarding conservation, and greater clarity in property law and regulation. But confronting these obstacles directly requires more political will than most westerners can presently muster.

California exemplifies the tension between the facts of groundwater depletion and the realities of water politics. Having long avoided the regulation of groundwater, and deep in the throes of an unprecedented drought, it enacted the Sustainable Groundwater Management Act in 2014, committing itself to reduce groundwater pumping to sustainable levels. Under the act, local entities are charged with developing groundwater sustainability plans and enforcing them; if they shrink from these duties, the state will step in and do so. This commitment is necessary and laudable, but the act contains a revealing omission: it explicitly avoids modifying or clarifying the rights to California groundwater.[11] Can a state achieve such a breakthrough in water policy without breaking (and remaking) its water law? Colorado and Kansas certainly hope so: their recent water plans brim with policy ambition but contain relatively little substantive legal reform.[12]

That imbalance stems from the fact that effective groundwater management requires the capability to resolve issues that are inchoate, diffuse, and frequently abstract. Does groundwater left in place have a value? What is the compensation required—if any—to reduce groundwater pumping? Where is the boundary between the private property right to use water and water's unquestionable status and value as a public and environmental resource? From a distance, these issues can polarize and frustrate further inquiry. The sheer scale of the West's groundwater crisis distorts our perception, creating that odd sort of respect that rationalizes inaction, like General Zebulon Pike and his apparently unclimbable peak.[13] On the ground, at the site- and case-specific level, these questions become the stuff of lawsuits between individual water rights owners, and of general adjudications to determine the water rights in a particular basin. These legal actions are decisive and conclusive, but they are typically expensive and slow, and can have uncertain results.[14] In short, the obstacles to effective groundwater management present us with problems of both appearance and scale. At the level of state policy and legislation, they seem too large and too abstract to resolve, while at the granular level of individual litigation and basin adjudication, definition and clarity arrive too slowly and at too high a cost.

This chapter proposes a different approach to these problems. It assumes that they are largely irresolvable by categorical legislative, administrative, or judicial action. Moreover, it concedes that these problems will likely never be fully resolved, owing to the nature of legal controversy, the hydrological variety of western watersheds, and the political diversity of western water management regimes. Instead, it begins with a modest question: what if we could better estimate and analyze how western aquifers respond to the impacts of pumping and irrigation? If we could do that, we could better determine both the quantities and characteristics of those water supplies. Knowing these things would go a long way toward reducing the uncertainties that undercut the will to confront the groundwater crisis and paralyze efforts toward effective reform. Most importantly, we could better comprehend the consequences of our management decisions.

Fortunately, recent innovations in hydrogeological analysis (HGA) have significantly improved hydrologists' understanding of the dynamics of groundwater systems. This improved understanding can provide a nonpolitical way for water users, regulators, and the public to assess and address the West's groundwater crisis together. The first part of the chapter summarizes the most forbidding obstacles to effective groundwater management in the West, primarily from a historical and legal perspective. With those obstacles in mind, the second part provides a summary of how HGA can significantly reduce them, but without requiring difficult changes in western water law. The third part concludes with some recommendations to incorporate HGA into state water management policy.

The Obstacles to Necessary Groundwater Management

Effective reform in western groundwater law and policy faces several entrenched obstacles. The property right in western groundwater is a confused property interest, unevenly grafted onto pre-existing doctrines. Unlike the statewide, basin-based management of surface water rights, management of groundwater rights typically takes place at the local level, through governance structures and powerful stakeholder groups that do not necessarily recognize or answer to statewide public interests. In the face of these obstacles, state leaders have tread lightly into the groundwater domain, but that deference has ultimately made the long-term protection of groundwater rights and supplies even more difficult.

The Property Right in Groundwater

A water right is a property right—the right to use water. In the humid eastern United States, where crops grow without irrigation, the right to use water generally comes with ownership of the land where that water supply is located, whether in a stream, lake, or underground. The owners of such lands are entitled to the reasonable use of these supplies; they cannot unreasonably diminish streamflows or groundwater levels to the detriment of neighbors and downstream users. These rules have evolved over centuries in England and the United States, mostly in response to the demands for water power that arose during the Industrial Revolution.[15] At bottom, they are rules of reasonableness and of equity.

A western water right is the legal product of a vastly different and severe habitat. The West is above all a drier place, where precipitation is too sparse to sustain crops and cattle without irrigation and where the highly variable flows of rivers and streams are often distant from where people need water. As a consequence of these conditions, western water law broke away from its eastern precedents almost immediately, starting with the California Gold Rush in 1849 and the Colorado Gold Rush a decade later. By the 1880s the basic rules of western water law had become established by state constitutions, state and federal statutes, and common law.[16] These rules claim to be clear and straightforward. One who diverts water from a stream or other water body, and puts that water to a recognized beneficial use, appropriates that water, and thus obtains a permanent right to use it, provided that the water is available, and provided also that one continues to use it. This property right does not result passively, from the mere ownership of riparian land. Rather, it results from the industry and investment required to construct canals, sluices, and pipelines, or to drill wells, and by the appropriators putting others on notice that they are in fact diverting and using the water.

In times of shortage, most western water law codes do not look to reasonableness and equity. That is because the conditions of aridity rule out such recourse: if every appropriator shared the shortage, not one would have sufficient water for their needs. Therefore, western water law generally follows the rule of priority, where the first in time is first in right. For example, where two appropriators each claim a right to divert fifty units of water from a stream, but the stream in a dry year can yield only fifty-five units, the appropriator who began diverting and using water in 1860 (or the successor) is entitled to receive the full share of fifty units, while the appropriator whose diversion and use dates from 1870 receives only five. Equitable considerations concerning the value, the reasonableness, and other merits of the respective uses are mostly irrelevant. In contrast to its eastern counterpart, the western water right endows its owner with considerably more power to exclude others—provided, of course, that the right is prior to other rights.

For the first eight decades of western water law, these rules evolved according to the assumptions and operational realities of surface water supplies—rivers, streams, and reservoirs—because these were the predominant sources of western water. Western irrigators and government geologists knew a fair amount about western groundwater supplies by the turn of the twentieth century, but they lacked the technological ability to exploit them on a large scale, compared to the gravity-powered canal systems that irrigated from river systems such as the Rio Grande and the South Platte. As long as the vast groundwater supplies of the West remained largely untapped, the distinct properties of groundwater did not pose much of a challenge to western water law codes.[17]

By the 1940s, however, American agriculture had begun to exploit the most important technological breakthroughs in the history of American irrigation: the high-capacity centrifugal pump and the rolling sprinkler system, which together could irrigate just about any farmland that rested above an aquifer. Between the 1940s and the 1970s, the pump and center pivot transformed western agriculture, bringing millions of dryland acres into irrigated production.[18] As a result, western legislatures wrestled with difficult legal and policy issues. Policymakers sought to balance the opportunity to put vast groundwater resources to beneficial use with the need to protect senior water rights, which suffered when excessive groundwater pumping lowered stream levels and water tables.

The groundwater revolution considerably complicated the legal landscape of western water. Doctrines became more diverse. Most western states generally extended the prior appropriation doctrine to groundwater but modified it to accommodate new development: new wells could henceforth lower water tables, but not unreasonably so.[19] Other states—including the top two states by irrigated acreage, California and Nebraska—did not adopt prior appropriation for groundwater. Instead, they retained the established doctrine of reasonable use, or developed the related doctrine of correlative rights, by which landowners receive water-use rights proportionate to their respective acreages.[20] Arizona retained the antiquated notion of groundwater "subflow," while Texas stuck steadfastly to the rule of capture.[21] These doctrinal complications accompanied new classifications of groundwater, depending on the hydrological relationship between groundwater and surface water and the historical relationship between surface water rights and groundwater rights.[22] By the 1970s, western water law, which Wallace Stegner had concluded was "so complex as to be utterly confusing to the layman," had become even more so.[23]

The groundwater revolution also upended the governance structures of western water. Surface water rights across the West remained under the jurisdiction of state engineers and state water boards—state agencies long deemed necessary to regulate river basins and their large and interdependent irrigation ditches, and to protect the public's interest in its water supplies. As Elwood Mead wrote over a century ago, "The impossibility of agreeing among themselves over the division of the stream, and the uncertainty and anxiety of the irrigators under the lower ditches, created a sentiment in favor of public supervision."[24] By contrast, the regulation of groundwater permitting and pumping increasingly fell within the jurisdiction of local districts, whose control was often exclusive, as in California and Nebraska.[25] Unlike state engineers and water boards, most local groundwater districts do not answer to the public but to their membership, which is typically restricted to landowners and holders of groundwater rights.[26] Groundwater irrigation across the West has consequently reduced states' effective authority over a substantial, even predominant, portion of their water resources—with a commensurate diminution of the public's role in the management of its groundwater.

The Insufficient Protection and Regulation of the Property Right in Groundwater

These complications in law and governance created a common consequence: to frustrate the protection and regulation of the property right in western water. The extension of the prior appropriation doctrine to groundwater exposed basic problems within that doctrine. Because it typically recognized a property right without due regard for the accuracy of the claim or the availability of the water, the doctrine had long been criticized for promoting excessive claims.[27] Yet groundwater, especially the vast but largely nonrenewable supplies of the Ogallala Aquifer, gave this problem a new twist. During the first decades of groundwater development, there appeared to be sufficient water supplies to satisfy most new claims; and that apparent availability effectively required state engineers to grant an increasing number of permits and rights, because these officials generally have a statutory duty to put unappropriated water to beneficial use.[28] As water levels began to decline during the 1970s, the problem of excessive claims returned, forcing state engineers and state legislatures to wrestle with it. Yet they continued to promote the development and use of groundwater supplies, often at the expense of senior rights and the water table.

Across the West, state engineers generally did not enforce or underenforce priorities, finding justification in statutes, administrative regulations, and court opinions that made it more difficult for holders of senior water rights to enforce them strictly and quickly.[29] This tacit approval of groundwater overpumping eventually forced interstate water litigation. New Mexico did not enforce priorities in groundwater pumping, which depleted streamflow in the Pecos River and drove Texas to file suit to enforce the Pecos River Compact in the U.S. Supreme Court in 1974.[30] Colorado similarly did not enforce priorities or sufficiently limit groundwater pumping in its portion of the Arkansas River basin, depleting stateline flows and leading Kansas to file suit to enforce the Arkansas River Compact in 1985.[31] Nebraska's repeated overpumping across the Republican River basin forced Kansas to sue to enforce the Republican River Compact twice, in 1998 and in 2010.[32] In each case the Supreme Court found that the interstate compacts included groundwater.[33]

Groundwater has presented other problems of underregulation as well. Irrigators and regulators in reasonable use/correlative rights states such as California and Nebraska have struggled with a basic question: what constitutes a reasonable use of water? The answer can be complicated. Our understanding of what is reasonable and what is wasteful changes over time. Modern irrigation technology and advances in crop genetics have enabled irrigators to raise yields with the same amount of water, or to maintain yields with less water. Regulators have embraced this progress as justification for reducing water allocations, based on the long-established precept that no one can obtain the right to waste water.[34] Yet equity and reasonableness cannot be confined to the amount of water used; they also consider the economic value of the use.[35] For example, nut trees in the Central Valley of California require large amounts of water, as do pasture crops used to feed beef and dairy cattle.[36] Unlike annual feed, grain, and vegetable crops, whose fields can be fallowed in dry years, trees must receive water every year to survive, placing a hard floor on the water savings that conservation can produce.[37] Yet offsetting this continuous thirst is a higher-value crop.

Consider a typical situation in the Central Valley. One farm raises almonds and pistachio nuts, which are processed and packaged locally, adding to their value, and then exported worldwide. A nearby farm raises alfalfa, clover, and regular hay, which feed local dairy cattle; while these crops can consume the same amount of water, they typically gross less money and generate fewer secondary economic effects and tax revenues.

In the event of a water shortage as dire as California's, should the shortage be borne according to water usage? Acreage? Crop yield? The gross or net incomes of the farms? Making a regulatory choice in such a situation can require the regulator to question a farmer's cropping decisions, and that is a perilous and litigious exercise.

These are just the major doctrinal problems; other problems abound. Legislative and common law classifications of groundwater supplies, as well as the creation and delineation of special groundwater management areas, have created numerous and inevitable conflicts because boundary lines have to be drawn somewhere, usually by some combination of hydrological reality and political necessity. Colorado has wrestled with its classification and management of groundwater basins, whether in the high-value aquifers that supply water to many of its Front Range suburbs or in the lower-value, primarily agricultural, designated basins above the Ogallala in the eastern third of the state.[38] Alongside these problems of classification are those of jurisdiction and control: when is it the duty of a local groundwater district to regulate groundwater and when is it the duty of the state? This is a challenge for Nebraska, which has the duty to limit its water use to comply with interstate compacts and decrees but has delegated almost all of its regulatory control over groundwater—by far its largest source of water—to local natural resources districts, which have avoided reductions in pumping.[39]

These varieties of doctrine, classification, and jurisdiction have produced complicated property rights in groundwater. Unfortunately, regulatory attention to these complications has too often come at the expense of the water supplies on which these rights depend. State engineers have a general duty to conserve the water resources of their state, but they also have the duty to put water to beneficial use, mostly by granting water rights.[40] But there is no explicit (and logically commensurate) duty to regulate water rights at a sustainable level. Indeed, where regulators take action to reduce groundwater pumping to sustainable levels, such action is extraordinary, exceptional, and deeply unpopular.[41] The proper administration of groundwater rights is more challenging than administering surface rights because determining well-to-well impairment takes time and expertise, especially regarding nonrenewable groundwater supplies.[42]

Regulations that govern groundwater pumping are susceptible to agency capture by irrigation interests, especially at the local level; such regulations can even subvert the intent of the statutes under which they are established. Across the Ogallala states, the unfortunate result of many local regulations and management plans is to preserve pumping at existing, often excessive, levels, thereby worsening the problem of groundwater depletion.[43] At some point, such regulatory failure can damage senior water rights to the point where their owners abandon the regulatory protections afforded them at the agency level and seek recourse in the courts instead.[44]

Resistance to Reform

For the past fifty years, the law of western groundwater has evolved alongside the regulation (and underregulation) of the groundwater right in practice. Some conclusions now seem clear—and forbidding. First, the property right in western groundwater has become problematic. Its legal definitions of authorized quantities and pumping rates are often deeply at odds with hydrological reality, and the legal protections the right affords are not the protections its holders want. Second, the economic motivations of groundwater irrigation appear to conflict with the long-term husbandry of the resource. Irrigators and their bankers often conduct their business with the unspoken belief that the present value of money exceeds the future value of water, even where groundwater supplies are dwindling rapidly and permanently. Finally, the political dominance of groundwater irrigation interests has generally militated against the deployment of effective strategies for and commitments to reducing groundwater depletion.

These have all informed the political culture of western groundwater, which has become understandably averse to some of the core tenets of western water law.[45] Groundwater "stakeholders" resist the tools of prior appropriation, which they view as a crude and heavy-handed regulatory tool rather than a means to protect senior rights. And they have a point, because so many groundwater supplies are severely overappropriated.[46] Irrigators can be hesitant about using lawyers and the legal process for protecting their rights, whether at the transactional, regulatory, or litigation level.[47] For the most part, they prefer the present ambiguities and confusions that have muddied the regulatory waterscape over the past several decades to the clarity of rights that legal decisions and adjudications usually bring, but at the expense of considerable time and money. Given a choice, they have acted rationally, using their political and regulatory influence to purchase delays in regulation instead.

As a consequence of all of this, the prospects of fundamental policy reform are not bright. Experts differ on the merits and demerits of the various legal and governance regimes for western groundwater, but they agree that property rights secured within one system become reliably resistant to redefinition.[48] Property rights in western groundwater will probably not change any time soon. From a practical standpoint, we need to develop different tools that will address groundwater depletion within these regimes. This is where hydrogeological analysis comes into play.

Hydrogeological Analysis of Groundwater Systems

Hydrogeological analysis is the use of an aquifer's response to groundwater pumping to ascertain the degree of aquifer "overdraft" and, by extension, the pumping reductions that would be required to reach a more sustainable level of withdrawals. HGA can be implemented in a variety of configurations, all of which are firmly rooted in the basic math of the aquifer water balance:

$$\text{Change in volume of water in aquifer} =$$
$$\text{Inflows into aquifer} - \text{Outflows from aquifer.}$$

When aquifer outflows (primarily from groundwater pumping) exceed inflows, as is typical in the West, the volume of water in the aquifer decreases. HGA, however, demands more than just a sound theoretical foundation; data quantifying aquifer conditions are also critical. In particular, data on water-level changes and groundwater pumping are essential if HGA is to move beyond the realm of an interesting mathematical exercise.

The three primary implementations of HGA, which differ according to their temporal and spatial scales, underlying assumptions, and fundamental objectives, are briefly outlined in the following subsections.

Impairment Analysis

Impairment analysis, the most common form of HGA, has been implemented across the West for decades to determine whether a proposed new water right, or a change to an existing right, would unreasonably affect, or "impair," the ability of nearby rights to obtain water. The analysis involves relatively simple mathematical expressions for the pumping-induced change in water level in a well (known as "drawdown") produced by the proposed right or the proposed change to an existing right. These expressions can be traced back to a model developed in 1935 for calculating the drawdown produced by pumping in an aquifer of infinite extent.[49] Information about the transmissive and storage characteristics of the aquifer is assumed to be known, although in reality, that may only be true in a rather approximate sense.

Several factors undercut the effectiveness of impairment analyses. First, the most common objective is to calculate the drawdown produced by pumping the proposed well over a single irrigation season. A standard assumption is that drawdown does not accumulate across irrigation seasons, which typically flies in the face of hydrological reality.[50] Second, the analyses emphasize discrete well-to-well interactions and so neglect the condition of the wider aquifer area. Third, the approach often gets bogged down in discussions of what level of drawdown constitutes "affecting" or "impairing" the ability of existing rights to obtain water. The result is that impairment analyses tend to underestimate the overall depletion problem and can lead to the granting of even more water rights, further exacerbating water shortages.

Safe Yield Analysis

Safe yield analysis, the second most common form of HGA, is a broader analysis directed at determining whether the aquifer is being "overtaxed" in a particular area. The procedure uses the water-balance equation with a defined recharge to calculate the pumping that would produce either stable water levels or, when that is deemed infeasible, an acceptable level of annual decline.[51] An estimate of recharge, often a highly uncertain quantity, is required, which can raise doubts about the reliability of the analysis.

A promising data-driven modification of this approach has recently been developed to overcome this limitation.[52] The modification involves rewriting the water-balance equation as

Change in volume of water in aquifer =
Net inflow − Groundwater pumping,

where net inflow equals the inflows into the aquifer minus all outflows from the aquifer except for pumping. The major advantage of this form of the water-balance equation is that it can be rearranged, as shown below, to directly estimate the amount of water flowing into the aquifer that is available for capture by pumping wells, the key quantity for determining prospects for aquifer sustainability.

We can rewrite this equation using standard groundwater notation to obtain

$$\Delta WL * Area * S_{aq} = I - Q,$$

where ΔWL is the average water-level change over an aquifer area for a given time interval, described in units of length; $Area$ is the aquifer area under consideration, described in units of area; S_{aq} is the volume of water released from aquifer storage by a given change in water level (typically one foot), a unitless quantity; I is the net inflow, described in units of volume; and Q is the total pumping, described in units of volume.

The equation can be further simplified by dividing both sides by $Area * S_{aq}$:

$$\Delta WL = \frac{I_{ua}}{S_{aq}} - \frac{Q}{Area * S_{aq}} \approx b - aQ,$$

where I_{ua} is net inflow per aquifer area ($= I/Area$), a and b are constants ($= 1/[Area * S_{aq}]$ and I_{ua}/S_{aq}, respectively), and all quantities are defined on an annual basis. The major assumption required to move to the right-hand side of the equation is that S_{aq} and I_{ua} are constant in time, which appears appropriate on the spatial scale of typical analyses (tens of square miles or larger) for the portions of the Ogallala Aquifer in Kansas.[53] The result is that a plot of ΔWL versus Q should be a straight line with a slope equal to a and an intercept equal to b. These slope and intercept values can then be used to develop important insights about current conditions in an aquifer and its likely future path.

A reasonable estimate of the average annual water-level change over an area (ΔWL) can be obtained from water-level measurements taken during the winter, when irrigation wells are typically not active across the Ogallala region.[54] If, as in northwestern Kansas, all high-capacity pumping wells are equipped with totalizing flowmeters, then a reasonable estimate of the annual water use (Q) can also be obtained. ΔWL can then be plotted versus Q and a straight line fit to that plot (see figure 6-1). The pumping that would produce stable water levels (Q_{stable}) can be calculated from the preceding equation by setting ΔWL to zero and rearranging the equation:[55]

$$Q_{stable} = \frac{b}{a} = I_{ua} * Area,$$

where $I_{ua} * Area$ can be considered equivalent to the recharge that is assumed to be known in traditional safe yield analyses but in this case is determined directly from the water-level and pumping data.

If the pumping reductions required to achieve Q_{stable} are judged to be economically or politically impractical, then a goal for the annual water-level change (ΔWL_{goal}) can be set and the pumping that would produce that change (Q_{goal}) can be calculated:

$$Q_{goal} = \frac{b}{a} - \frac{\Delta WL_{goal}}{a}.$$

As with the impairment analysis, this approach can get bogged down in discussions of what level of annual decline is acceptable, that is, what would constitute "overtaxing" the aquifer.

The major limitation of this approach is its dependence on the quantity and quality of available water-level and pumping data. Although such data are the most critical elements needed for any type of aquifer assessment, no concerted effort has been made to acquire them on a consistent basis across the West. The result is that all too often, and particularly with respect to pumping data, we are forced to deal with highly uncertain quantities, such as annual water use estimated from the electricity usage of irrigation pumps.[56] As more attention is given to acquiring such data on a routine basis, the modified form of the safe yield analysis should become a powerful analytical and policy tool for rapid assessment of an aquifer's prospects for sustainability and the impact of proposed management activities. This theoretically sound, data-driven approach is particularly well suited for use in seasonally pumped aquifers that support irrigated agriculture in semiarid settings.

Distributed Parameter Groundwater Flow Model Analysis

This third and increasingly common form of HGA is typically performed to determine an aquifer's response to groundwater development over the scale of the entire aquifer or a significant portion of it. In a distributed parameter groundwater flow model, an aquifer is subdivided into a network of cells; the aquifer area represented by a single cell depends on the desired spatial resolution of the analysis. (The area of a typical cell ranges from tens of thousands of square feet to a few square miles.) The water-balance equation is applied to each cell, and the cells are linked to allow for the exchange of water across the network. This approach is particularly well suited for application over relatively large spatial (hundreds of square miles and larger) and temporal (multiyear) scales. It is commonly used to assess the aquifer-wide impact of proposed wells, the future declines produced by different management strategies, and the estimated usable lifetime of the groundwater supply for a given management strategy.[57]

The fundamental factors limiting the effectiveness of distributed parameter modeling are again the data requirements. Water-level and pumping data are needed, as is information about recharge and the transmissive and storage characteristics of the aquifer at the scale of individual model cells. Model input requirements far exceed available data, so hydrologists use a procedure known as calibration to obtain estimates of the unknown quantities. Calibration involves adjusting these estimates until an acceptable level of agreement is obtained between existing water-level data and the model-calculated water levels.[58] Although the technical sophistication of distributed parameter modeling can imbue the analysis with an aura of high reliability, this approach is even more heavily dependent on the quality and quantity of available data than the other forms of HGA. If data are sparse, the reliability of the analysis will be questionable, regardless of the particulars of the calibration procedure.

Clarifying the Current Realities of the Water Right with Hydrogeological Analysis

HGA results should not be considered to be the final, time-invariant quantification of aquifer conditions, as the elements of an aquifer's water balance are prone to change in the semiarid and arid West, particularly in the face of continuing climate change. The available extractable water, and thus the quantity available to satisfy groundwater rights, have decreased substantially and will continue to decrease with time in these settings. HGA presents a scientifically defensible means to quantify those changes.

As an example, the modified safe yield analysis can be applied to nineteen years of data (1996–2014) from Groundwater Management District 4 (GMD 4) in northwestern Kansas. Figure 6-1 is a plot of ΔWL versus Q for this period. ΔWL is the average for the 182 wells measured every year from 1996–2015 and Q is the sum of reported use from a maximum number of 4,211 pumping wells (this number varies slightly from year to year).[59] The slope (a) and intercept (b) of the best-fit line yield a Q_{stable} of 336,000 acre-feet, which equates to an average I_{ua} of 1.3 inches per year over the 3.12 million acre district.[60] Given the average annual water use (Q) from 1996 to 2014 of 430,000 acre-feet, a reduction in average annual use of approximately 22 percent would produce stable water levels. Clearly, the aquifer in this region is being overtaxed at the scale of the district as a whole. Equally clear, however, is that stable water levels, at least in an average sense, may not be a pipe dream for the short term. The clustering of points about the best-fit line in figure 6-1 (see position of the 2014 point) indicates that there is little temporal variability in S_{aq} and I_{ua}

FIGURE 6-1

Average Annual Water Level Change versus Annual Water Use Plot for Nineteen Years of Data from Groundwater Management District 4 (GMD 4), Kansas

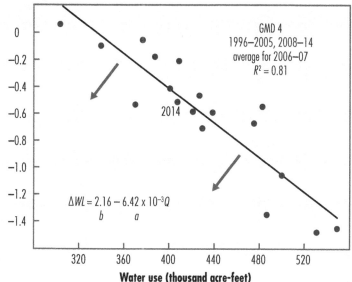

over the nineteen years of the analysis. However, net inflow is likely to diminish in the future, as infiltrating water resulting from past inefficient irrigation practices completes its journey to the water table. Repeating the analysis every four to five years will reveal how I_{ua}, and thus Q_{stable}, are changing. Decreases in these quantities will produce a downward shift in the best-fit line in the direction of the arrows in figure 6-1, an indication of the need to revisit the Q_{stable} calculations. Similar analyses can be performed for smaller areas to assess conditions on scales more pertinent to individual groundwater rights.[61] Regardless of the scale of the analysis, the impact of climate change on aquifer conditions should be recognizable through changes in Q_{stable} with time.

As shown by this example, HGA presents a promising means to clarify the property right in water and to improve its regulation and adjudication. In particular, HGA provides metrics, such as Q_{stable} or Q_{goal}, that can define the current realities of a right. This recognition of the physical constraints imposed by hydrogeological conditions would likely improve the process by which the regulation and administration of water rights take place.

The Policy Promise of Hydrogeological Analysis

As shown in the preceding section, HGA provides a better understanding of the characteristics of groundwater, both in time and in place. It can produce a relatively accurate "water budget" for a particular area, thereby quantifying the realities of an area's water rights. More important, however, HGA can capture the dynamic effects of groundwater rights on one another, and on the groundwater supply in an aquifer, under various management scenarios. Because it can perform these functions, HGA holds great promise to improve groundwater management in at least four ways.

First, HGA enables better protection of property rights in groundwater over the long term. The reason for that is straightforward: holders of water rights cannot protect what they cannot define. The last two forms of HGA discussed here, modified safe yield analysis and distributed parameter groundwater flow model analysis, are promising means to clarify the property right in water and thus to improve its regulation. In the semiarid and arid West, the physics of the aquifer water balance constrain a property right in groundwater to be a function of time, because an increasingly overtaxed aquifer will yield a diminishing amount of water to a well over time. HGA provides metrics that can define the current realities of a water right, and thus allows policymakers to begin to confront, address, and potentially alleviate the overappropriation gap— where the "paper water" legally granted to water rights holders significantly exceeds the "wet water" that is actually available to supply those rights. Because HGA provides a specific and dependable means of predicting the consequences of regulatory action, HGA-based administration of water rights can be performed quickly; it can be performed iteratively and so improved over time; and its consequences can be predicted with a relatively high degree of confidence. These are substantial improvements over the current regulatory situation for most of western groundwater, where regulatory consequences are slow and uncertain, and can produce paralysis as a consequence. In addition, HGA enables owners, regulators, and policymakers to consider explicitly the impacts of groundwater pumping on surface water rights and wildlife habitats, and on groundwater-dependent ecosystems.

Second, HGA enables policymakers to manage groundwater resources more clearly. Because HGA can produce relatively accurate forecasts of groundwater supplies and pumping impacts under various management scenarios, holders of groundwater rights can make management decisions with a much higher degree of predictability, enhancing their role in the management process. At the group level, the modeling of pumping impacts has made it possible for irrigators farming ninety-nine square miles of ground in Kansas GMD 4 to reduce groundwater pumping by 20 percent over a five-year period. They were able to take this conservation step largely because HGA helped assure them that they would benefit from those reductions over the long term.[62] At the individual level, HGA can serve a similar function. The argument for groundwater conservation becomes much stronger when HGA can demonstrate that reducing the present pumping from a particular well will conserve the groundwater that supplies it, enabling a longer duration of beneficial use in the future. By demonstrating that individual water rights holders can harvest their own water savings, and not lose them to other rights, HGA can be an effective educational tool for reducing stakeholder resistance to policy reforms.

This clarity can be achieved across the various legal regimes for western groundwater. In states that follow the prior appropriation doctrine, irrigators across the Ogallala have generally resisted regulatory initiatives to reduce pumping for three reasons. First, junior rights far outnumber senior rights. Second, protecting one senior right from impairment can require dramatic reductions in pumping from numerous nearby wells. Finally and consequently, many groundwater irrigators tend to believe that the prior appropriation doctrine is arbitrary and unfair. HGA may even increase skepticism about the doctrine—especially where it demonstrates that the most important attribute of a well may not be its temporal priority in the past but its hydrological capacity in the future. Nonetheless, owners of senior rights are not about to surrender their priorities. The ability of HGA to forecast consequences relatively accurately gives it great potential to help navigate the difficult path between the legal clarity of the prior appropriation doctrine on one side and its unpleasant consequences on the other—because HGA makes those consequences much more predictable. Rather than await a water rights impairment investigation, irrigators can take their own management initiatives. The best example of this potential is again in that ninety-nine square mile area in Kansas GMD 4. Irrigators there not only accepted the 20 percent reduction in pumping; they applied it to all irrigators regardless of priority, largely because HGA helped them recognize that the reductions would affect them similarly. This recognition was critical in enabling the irrigators to achieve consensus on the water conservation plan.[63] For reasonable use/correlative rights states such as California and Nebraska, HGA similarly provides a data-driven means of confronting the question of how much water usage is reasonable by allowing groundwater irrigators to understand the long-term consequences of different levels of pumping. HGA can thus serve to displace arcane questions of legal equity to the background—where they probably belong.

At the larger political level, the clarity of these consequences will force people to come to terms with them—and, because of the politics of groundwater, that is progress indeed. It will motivate the public, legislatures, and stakeholders to determine water boundaries more clearly, regardless of doctrine or jurisdiction. The two most important decisions are those concerning the boundary between renewable and nonrenewable groundwater resources and the boundary between the private right to pump groundwater and the public's claim to groundwater as a public resource. HGA enables a candid discussion of the consequences of where these boundaries are drawn—a discussion that was not possible a generation ago and that has too often descended into legal, doctrinal, and regulatory confusion largely as a consequence.

Third, HGA can significantly improve markets for groundwater. HGA enhances the marketability of the property right in groundwater by clarifying its current "value" and defining a transparent and predictable means of reevaluating that value on a set timetable. Such clarity and transparency should enable the development of more sophisticated water rights transactions—which can help to address problems of over-appropriation and misdistribution. One such transaction is known as a "wheeling agreement," in which owners of agricultural land with multiple water rights contract to transfer their respective rights to a different user, typically a municipal or industrial one, but on a rotating basis. Such agreements can have the benefit of avoiding the permanent dry-up of ir-rigated land—the so-called "buy and dry" problem—and are gaining support across the West.[64] Another water market mechanism is that of the water bank, whereby irrigators "deposit" unused water from their annual water rights and make it available for "withdrawals" by other nearby irrigators. These types of transactions can be vexed with regula-tory complications and delays because they involve a potentially complex combination of hydrological impacts as water is moved to and from dif-ferent places of use. HGA can significantly reduce such transactional and regulatory friction.

Finally, HGA holds significant promise as a tool that can reduce the scale, the burdens, and perhaps even the intensity of groundwater-related regulatory and legal disputes. In the regulatory domain, HGA can pro-vide an improved understanding of what is possible from a regulatory standpoint and, just as important, what is not. For example, in the Rio Grande basin in southern Colorado, rules governing groundwater usage now depend significantly on HGA, specifically, the Rio Grande Decision Support System, which employs a distributed parameter groundwater model to simulate groundwater flows and estimate pumping impacts.[65] In northwestern Kansas, GMD 4 used a similar model to develop a local groundwater conservation plan.[66] In the legal domain, HGA can provide a stipulated record of the physics-based contours of the property right in groundwater in time and space. Because much of the time and expense of water rights litigation are dedicated to a "war of the experts," such stipu-lation should help overcome the traditional and justified objections to lawsuits as overly expensive, time-consuming, and uncertain proceedings that produce fundamentally unsatisfactory results. At the basinwide scale of water rights adjudications, HGA holds significant promise as a tool that can efficiently and effectively confront the legal and policy problems of overappropriation of regional aquifers.[67]

As policymakers, regulators, courts, and individual litigants struggle with the many problems of permanent groundwater depletion, it is nearly certain that forced reductions in pumping will give rise to the claim that they amount to an unconstitutional regulatory taking of property rights. In regard to this claim, HGA will prove to be a most valuable and potentially decisive tool. By showing how significant but tolerable pumping reductions imposed in the present can extend the practical life of groundwater rights well into the future—especially in largely nonrenewable aquifers such as the Ogallala Aquifer—HGA can provide a compelling case for such regulation, buttressing traditional legal arguments in support of such reductions.[68]

Conclusion

Many of the West's most difficult natural resource issues are problems about how the past relates to the present, and the challenge of groundwater depletion is no exception. The legal and policy doctrines undergirding the property right in water, its proper regulation, and the public interest in water supply are founded on nineteenth- and early twentieth-century assumptions. As the West has changed, especially over the past fifty years, many of those assumptions have evolved, and water law has evolved accordingly. For example, the law once considered letting water run downstream to be a waste of water; but such "in-stream" flows are now recognized across the West as a valuable beneficial use, in the form of recreational water rights for fishing, aquatic habitat, and boating.[69] Champions of the prior appropriation doctrine point to this sort of adaptation as evidence of the doctrine's continuing vitality and merit.[70] Critics of prior appropriation and other long-established natural resources doctrines, on the other hand, emphasize their anachronism; to them, such doctrines are the "lords of yesterday," exercising an inappropriate and disproportionate power in the modern West.[71] This is not just an academic debate because property rights in water, like all property rights regimes, fundamentally depend on long-established doctrines.[72] For better and for worse, holders of western water rights have relied on them for more than 150 years. That is considerable legal and historical momentum.

Yet regardless of one's position, there is a limit to how far these doctrines can evolve, and how flexibly they can be applied. At some point, efforts to retain them come at the expense of common sense—and that is especially the case with new technologies of natural resources extraction. Does the extraction and disposal of water as part of coal bed methane production—water that operators consider to be a nuisance and do not use—nonetheless count as a "use" of water?[73] Do the ancient doctrines of trespass and the rule of capture apply to hydraulic fracturing for oil and gas, which can take place more than two miles below the surface of the Earth?[74] Learned courts are struggling with these and other questions, and their opinions are appropriately fractured, giving rise to the recognition that modern natural resources issues require a modern property analysis.[75] As these new technologies have come to prominence, reliance on antiquated notions of natural resources extraction has itself become less reliable.

This is probably where we are with groundwater. As we have seen, the groundwater revolution was a technological disruption of the first order, upending long-established assumptions about water availability and use across the West. That disruption had important political and cultural consequences, as groundwater irrigators and other major water users established distinct governance structures that reflected their own beliefs about the proper regulation of groundwater resources.[76] Holders of groundwater rights have depended on those rights for several generations, but their wells are becoming less and less dependable. As we look toward a future of permanent depletion and continuing climate change, both the distant and the relatively recent past seem especially problematic. What do we need to escape from this thicket of tangled doctrine, history, and policy?

More than anything else, we need facts. By providing a detailed factual picture of western aquifer systems and the present and future impacts of groundwater pumping on those systems, HGA is the most important tool we can employ to understand how property rights in groundwater work in practice. That understanding is the necessary first step in any proper reevaluation of western groundwater law and policy, and it is also our best hope for reducing the size and severity of legal and political conflicts over our aquifers. The more we can rely on the facts and uncontroversial forecasts of HGA, the less we will need to resort to strained and distant legal doctrines, and the better prepared we will be to confront the realities of the future through necessary policy reforms.

Notes

The authors gratefully acknowledge the review comments provided by Rex Buchanan and Gil Zemansky. The work of the second author was supported in part by the National Science Foundation (NSF) under award 1039247 and the U.S. Department of Agriculture (USDA) under subaward RC104693B. Any opinions, findings, and conclusions or recommendations expressed in this material are those of the authors and do not necessarily reflect the views of the NSF or USDA.

1. Virginia L. McGuire, "Water-Level Changes in the High Plains Aquifer, Predevelopment to 2007, 2005–06, and 2006–07," *U.S. Geological Survey Scientific Investigation Rept. 2009-5019*, 2009, p. 1.

2. Leonard F. Konikow, *Groundwater Depletion in the United States (1900–2008)* (U.S. Geological Survey, 2013), p. 5. In deference to common usage, this chapter refers to the High Plains–Ogallala Aquifer as the Ogallala Aquifer, or just the Ogallala.

3. Stephanie L. Castle, Brian F. Thomas, John T. Reager, Matthew Rodell, Sean C. Swenson, and James S. Famiglietti, "Groundwater Depletion during Drought Threatens Future Water Capacity of the Colorado River Basin," *Geophysical Research Letters* 41, no. 16 (2014), pp. 5904–11.

4. Sandra Postel, "Groundwater Depletion in Colorado River Basin Poses Big Risk to Water Security," *National Geographic Water Currents*, July 30, 2014.

5. Bettina Boxall, "Overpumping of Central Valley Groundwater Creating a Crisis, Experts Say," *Los Angeles Times*, March 18, 2015 (www.latimes.com/local/california/la-me-groundwater-20150318-story.html).

6. See, for example, Kan. Stat. Ann. § 82a-701(g) (2015) (defining a water right as a real property right, appurtenant to the land on which it is used, but severable from it).

7. Justin S. Mankin, Daniel Viviroli, Deepti Singh, Arjen Y. Hoekstra, and Noah S. Diffenbaugh, "The Potential for Snow to Supply Human Water Demand

in the Present and Future," *Environmental Research Letters* 10, no. 11 (2015), pp. 114016–25.

8. United States v. New Mexico, 438 U.S. 696 (1978) (describing the limits of the federal reserved water rights to which national forests are entitled).

9. See, for example, U.S. Forest Service, Notice of Withdrawal of Proposed Directive on Groundwater Resource Management, Forest Service Manual 2560, 80 Fed. Reg. 35,299 (June 19, 2015). The United States' federal landholdings in the West are over six times greater than the size of California (Bureau of Land Management, *Public Lands Statistics 2000* [Department of the Interior, 2000], table 1-3, p. 7). While national forests have a valid claim to federal reserved water rights, most lands managed by the Bureau of Land Management do not.

10. Most western water codes dedicate state water supplies to the public, subject to private appropriation in the form of water rights, which are use rights. See, for example, Kan. Stat. Ann. § 82a-702 (2014). Texas, which alone among the western states follows the rule of capture for groundwater, is also an outlier in this regard. The Texas Supreme Court recently held that landowners own the groundwater in place beneath their lands; as a consequence, regulating groundwater withdrawals through a permitting system constitutes a compensable regulatory taking. Edwards Aquifer Authority v. Day, 369 S.W.3d 814, 831 (Tex. 2012).

11. California Sustainable Groundwater Management Act, Cal. Water Code § 10720 *et seq.*, esp. § 10720.5 (West 2015).

12. Colorado Water Conservation Board, *Colorado Water Plan—Final 2015* (Denver, 2015) (http://coloradowaterplan.com/); Kansas Water Office, *A Long-Term Vision for the Future of Water Supply in Kansas* (Topeka, January 2015) (www.kwo.org /Vision/rpt_Kansas_Water_Vision_%20Final_%20Draft_%20012815.pdf).

13. *The Southwestern Journals of Zebulon Pike 1806–1807,* edited by Stephen Harding Hart and Archer Butler Hulbert (University of New Mexico Press, 2006), p. 144.

14. See, for example, Joseph M. Feller, "The Adjudication That Ate Arizona Water Law," *Arizona Law Review* 49 (2007), pp. 405, 406.

15. Martin J. Horwitz, *The Transformation of American Law, 1780–1860* (Harvard University Press, 1979), pp. 31–62.

16. For an introductory history of the development of western water law, see Donald J. Pisani, *To Reclaim a Divided West: Water, Law, and Public Policy, 1848–1902* (University of New Mexico Press, 1992).

17. See, for example, Erasmus Haworth, *Underground Waters of Southwestern Kansas,* edited by U.S. Geological Survey (U.S. Department of the Interior, 1897); George Evert Condra, *Geology and Water Resources of the Republican River Valley and Adjacent Areas, Nebraska,* U.S. Geological Water-Supply Paper 216 (U.S. Department of the Interior, 1907), pp. 31–39. Artesian wells, which are under sufficient pressure to bring groundwater to the surface without pumping, were a happy

exception, and western water codes governed their use accordingly: see, for example, Kan. Stat. Ann. § 42-330 to 42-332 (1891).

18. James Aucoin, *Water in Nebraska: Use, Politics, Policies* (University of Nebraska Press, 1984), pp. 38–41; D. E. Green, *Land of the Underground Rain: Irrigation on the Texas High Plains, 1910–1970* (University of Texas Press, 1973).

19. This principle was adopted by Colorado, Idaho, Kansas, Montana, Nevada, New Mexico, North Dakota, Oregon, South Dakota, Utah, Washington, and Wyoming. For a survey of groundwater rights across the western states, see Wells A. Hutchins, Harold H. Ellis, and J. Peter DeBraal, *Water Rights Laws in the Nineteen Western States*, 3 vols. (U.S. Department of Agriculture, 1972–77), 2:631–756, 3:141–648. In Colorado, senior rights to Ogallala supplies are no longer entitled to the protection of historical water levels. Colo. Rev. Stat. §§ 37-90-102(1), 37-90-103(6), 37-90-111(1)(a)-(b) (2014). Kansas made a similar accommodation by redefining the criteria for evaluating new water rights applications to include economic standards as well as hydrological ones. Kan. Stat. Ann. § 82a-711 (2015).

20. These variations were adopted by California, Nebraska, and Oklahoma. Hutchins and others, *Water Rights Laws in the Nineteen Western States*, 3:179–213, 332–65, 423–40.

21. Arizona has spent decades litigating what constitutes "subflow" in the Gila River basin; see Feller, "The Adjudication That Ate Arizona Water Law," pp. 423–25. For a recent discussion of the rule of capture as applied to Texas groundwater, see Edwards Aquifer Authority v. Day, 369 S.W.3d 814, 831 (Tex. 2012). See also Hutchins and others, *Water Rights in the Nineteen Western States*, 3:162–79, 503–35.

22. Colorado, for example, makes a fundamental distinction between groundwater that is tributary to natural streams and groundwater that is not. Colo. Rev. Stat. § 37-92-101 et seq. (statutory regime for tributary groundwater); *Id.*, § 37-90-101 et seq. (statutory regime for other types of groundwater, most importantly "designated" and "nontributary" groundwater, categories which include most of Colorado's Ogallala Aquifer supplies).

23. Wallace Stegner, *Beyond the Hundredth Meridian: John Wesley Powell and the Second Opening of the West* (New York: Houghton Mifflin, 1954; repr. New York: Penguin Books, 1992), p. 411, n. 9. Stegner underestimated the problem; he issued this opinion in 1954, at the dawn of the groundwater revolution.

24. Elwood Mead, *Irrigation Institutions: A Discussion of the Economic and Legal Questions Created by the Growth of Irrigated Agriculture in the West* (New York: Macmillan, 1903), p. 145.

25. Norris Hundley has described the local management of groundwater in California as "a chaotic and environmentally destructive practice of management by numerous local water districts and agencies—a practice that has meant no management at all." Hundley, *The Great Thirst: Californians and Water, 1770s–1990s* (University of California Press, 2001), p. 530. For a discussion of local control over

groundwater in Nebraska, see Spear T Ranch v. Knaub, 691 N.W.2d 116 (Neb. 2005).

26. In Kansas, for example, membership in groundwater management districts is limited to owners of forty or more contiguous rural acres and owners of groundwater rights. Kan. Stat. Ann. § 82a-1021(e) (2015).

27. Pisani, *To Reclaim a Divided West*, p. 37.

28. See, for example, Kan. Stat. Ann. § 82a-711(a) (2015).

29. See, for example, Fellhauer v. People, 447 P.2d 986 (Colo. 1968).

30. Texas v. New Mexico, No. 65 Orig. (1974–90): see especially 462 U.S. 554 (1983), 482 U.S. 124 (1987).

31. Kansas v. Colorado, No. 105 Orig. (1985–2009); see especially 533 U.S. 1 (2001).

32. Kansas v. Nebraska & Colorado, No. 126 Orig. (1998–2003, 2010–15); see especially Kansas v. Nebraska & Colorado, 135 S. Ct. 1042 (2015).

33. Kansas v. Nebraska & Colorado, First Report of the Special Master (Subject: Nebraska's Motion to Dismiss) (January 28, 2000), 23–31.

34. See, for example, Envtl. Defense Fund, Inc. v. E. Bay Mun. Util. Dist., 52 Cal. App. 3d 828 (Cal. 1975); City of Barstow v. Mojave Water Agency, 5 P.3d 853, 864 (Cal. 2000).

35. Colorado v. New Mexico, 459 U.S. 176, 183 (1982).

36. According to the Almond Board of California, almond trees use 9 percent of California's agricultural water—3.03 million acre-feet annually. Almond Board of California, "Get the Facts about Almonds and Water" (updated May 2016) (www.almonds .com/get-facts-about-almonds-and-water?gclid=CPvIltLss8kCFYk9gQodkqQJhw). According to the Glenn-Colusa Irrigation District, walnuts require 4.4 feet of water annually, while pasture crops require 4.9 feet annually. Glenn-Colusa Irrigation District, "2016 Water Application Instructions" (http://media.wix.com/ugd/c88b6b_a1b2b2dc 5d3a4825b7032b1c25d24066.pdf).

37. Felicity Barringer, "Water Source for Almonds in California May Run Dry," *New York Times*, December 27, 2014, p. A24.

38. Colorado Ground Water Commission v. North Kiowa Bijou Ground Water Management District, 77 P.3d 62, 70 (Colo. 2003).

39. Kansas v. Nebraska & Colorado, No. 126 Orig., 135 S. Ct. 1042 (2015).

40. See, for example, Kan. Stat. Ann. § 82a-706 (2015).

41. John C. Peck, "Groundwater Management in Kansas: A Brief History and Assessment," *Kansas Journal of Law & Public Policy* 15 (2006), p. 441.

42. See, for example, Kan. Admin. Regs. §§ 5-4-1, 5-4-1a (2015).

43. See, for example, Michael K. Ramsey, "Kansas Groundwater Management Districts: A Lawyer's Perspective," *Kansas Journal of Law & Public Policy* 15 (2006), pp. 517, 522.

44. Garetson Bros. v. American Warrior, Inc., 347 P.3d 687 (Kan. Ct. App. 2015).

45. Burke W. Griggs, "Irrigation Communities, Political Cultures, and the Public in the Age of Depletion," in *Bridging the Distance: Common Issues of the Rural West*, edited by David Danbom (University of Utah Press, 2015), pp. 101–47.

46. James J. Butler Jr., Randy L. Stotler, Donald O. Whittemore, and Edward C. Reboulet, "Interpretation of Water-Level Changes in the High Plains Aquifer in Western Kansas," *Groundwater* 51, no. 2 (2013), pp. 180–90.

47. This is a generalization, of course. Some states see more water rights–related litigation than others, especially where legal procedure is central to the regulatory process, as in Colorado's water court system, which performs rolling adjudications of individual water rights.

48. See, for example, Reed D. Benson, "Alive but Irrelevant: The Prior Appropriation Doctrine in Today's Western Water Law," *University of Colorado Law Review* 83 (2012), p. 675.

49. Charles V. Theis, "The Relation between the Lowering of the Piezometric Surface and the Rate and Duration of Discharge of a Well Using Ground-water Storage," *Transactions of the American Geophysical Union*, 16th Annual Meeting (1935), pt. 2, pp. 519–24.

50. See Butler and others, "Interpretation of Water-Level Changes in the High Plains Aquifer in Western Kansas," figure 2, for an example of an extreme case.

51. Recharge is defined as the processes involved in the addition of water to the aquifer; see the *Glossary of Hydrology*, edited by William E. Wilson and John E. Moore (Alexandria, Va.: American Geological Institute, 1998), p. 163. In the Ogallala Aquifer, the sources of recharge are primarily precipitation and excess irrigation water.

52. James J. Butler Jr., Donald O. Whittemore, and Blake B. Wilson, "A New Approach for Assessing the Future of Aquifers Supporting Irrigated Agriculture," *Geophysical Research Letters* 43, no. 5 (2016), pp. 2004–10.

53. Ibid., p. 2005.

54. Richard D. Miller, Rex C. Buchanan, and Liz Brosius, "Measuring Water Levels in Kansas," *Kansas Geological Survey Public Information Circular 12* (Lawrence, 1999); Geoffrey C. Bohling and Blake B. Wilson, "Statistical and Geostatistical Analysis of the Kansas High Plains Water-table Elevations, 2012 Measurement Campaign," *Kansas Geological Survey Open-File Report 2012–16* (Lawrence, Kansas, 2012).

55. Butler, Whittemore, and Wilson, "A New Approach for Assessing the Future of Aquifers Supporting Irrigated Agriculture," p. 2005.

56. The exception is Kansas, where every nondomestic groundwater right in the Ogallala is required to have a totalizing flowmeter and water rights holders are required to report the totals from their meters annually (Butler, Whittemore, and Wilson, "A New Approach for Assessing the Future of Aquifers Supporting Irrigated Agriculture," p. 2006).

57. See Mary P. Anderson, William W. Woessner, and Randall J. Hunt, *Applied Groundwater Modeling* (San Diego: Academic Press, 2015).

58. Ibid.

59. Heavy snows delayed the 2007 water-level measurements from early January to late February through early April, so the 2006 and 2007 water-level and water-use values are averaged and plotted as a single point.

60. Confidence intervals are not presented for this illustrative example. Coefficient of determination (R^2) in figure 6-1 indicates that 81 percent of the variation in ΔWL can be explained by variation in Q; the major sources of the variability about the best-fit line are likely measurement error and year-to-year variability in net inflow and in the distribution of pumping within an irrigation season.

61. As the spatial scale of the analysis decreases, a two- to three-year averaging window can help diminish the impact of measurement errors, isolated off-season pumping, and other factors that typically bedevil such analyses as the number of wells used for calculation of ΔWL falls into the lower single digits.

62. Kansas Department of Agriculture, Division of Water Resources, Order of Designation Approving the Sheridan 6 Local Enhanced Management Area within Groundwater Management District No. 4 (April 17, 2013), pp. 20–23 (http://dwr.kda.ks.gov/LEMAs/SD6/LEMA.SD6.OrderOfDesignation.20130417 .pdf).

63. Ibid., pp. 11–18.

64. The Arkansas Valley "Super Ditch" is one prominent example of a rotational water agreement between farms in the Arkansas Valley of eastern Colorado and suburban municipalities near Denver. See Tyler G. McMahon and Mark Griffin Smith, "The Arkansas Valley Super Ditch: A Local Response to 'Buy and Dry' in Colorado Water Markets," Colorado College Working Paper 2011-08 (September 5, 2011) (http://ssrn.com/abstract=1922444).

65. Colorado Department of Natural Resources, Division of Water Resources, Rules Governing the Withdrawal of Groundwater in Water Division No. 3 (the Rio Grande Basin) and Establishing Criteria for the Beginning and End of the Irrigation Season in Water Division No. 3 for all Irrigation Water Rights, Order of the State Engineer (September 23, 2015), 4.20–24.

66. Burke W. Griggs, "Beyond Drought: Water Rights in the Age of Permanent Depletion," Kansas Law Review 62 (2014), pp. 1291–92.

67. For a more extensive discussion of this potential, see Burke W. Griggs, "General Stream Adjudications as a Property and Regulatory Model for the Ogallala Aquifer, Wyoming Law Review 15 (2015), pp. 413–60.

68. See, for example, Dave Owen, "Taking Groundwater," Washington University Law Review 91 (2013), pp. 253, 254.

69. See, for example, Colo. Rev. Stat. § 37-92-102(3)-(6) (2015).

70. See, for example, Justice Gregory J. Hobbs Jr., "Priority: The Most Misunderstood Stick in the Bundle," Environmental Law 32 (2002), p. 37.

71. Charles Wilkinson, Crossing the Next Meridian: Land, Water, and the Future of the West (Washington, D.C.: Island Press, 1992), pp. 19–24.

72. Williams v. City of Wichita, 374 P.2d 578, 596-612 (Kan. 1962) (J. Schroeder, dissenting).

73. Vance v. Wolfe, 205 P.3d 1165 (Colo. 2009).

74. Coastal Oil & Gas Corp. v. Garza Energy Trust, 268 S.W.3d 1 (Tex. 2008).

75. David E. Pierce, "Carol Rose Comes to the Oil Patch: Modern Property Analysis Applied to Modern Reservoir Problems," *Penn State Environmental Law Review* 19 (2011), p. 241.

76. Griggs, "Irrigation Communities."

Chapter 7

Southeastern Florida
Ground Zero for Sea-Level Rise

DOUGLAS YODER

In 2007 the Organization for Economic Cooperation and Development
(OECD) published a report that ranked major cities worldwide in terms
of the current (2005 data) and future (2070 projection) coastal flooding
risks with respect to population and the value of assets.[1] Miami (including all of Miami-Dade County) ranked number nine in the world with respect to population at risk and number one in the world with respect to
the value of assets at risk. Southeastern Florida is no stranger to both the
risks and the rewards associated with climate and weather. A subtropical
setting moderated by ocean breezes, miles of white sandy beaches for residents and tourists to enjoy, and abundant high-quality water managed to
control flooding all contributed to a series of development booms, punc-
tuated at intervals by dramatic hurricanes and economic downturns. Historically, these recurring risks have been acknowledged, understood to
some extent, and mitigated through building code modifications, viable
property insurance, emergency response capability, and recovery planning.
The addition of climate change, and most particularly sea-level rise, to the
list of factors to be considered has introduced a new challenge to the region, one that requires a longer-term view and an iterative, adaptive

management approach to comprehensive planning and infrastructure investment, including water-supply and wastewater services infrastructure and operations. Past experience can no longer be the exclusive benchmark by which future conditions are forecast.

Regional Background

For purposes of this discussion, southeastern Florida is considered to include Palm Beach County (Palm Beach and West Palm Beach), Broward County (Fort Lauderdale), Miami-Dade County (Miami, Miami Beach, and Hialeah), and Monroe County (Key Largo and Key West). Together these counties include a land area of 5,000 square miles and a population of 5.8 million residents (about one-third of Florida's total population), clustered primarily on a series of barrier islands and adjacent shoreline stretching 200 miles from north to southwest. The region includes Everglades National Park to the west and Biscayne National Park to the southeast, emphasizing the historical importance of the natural conditions that prevailed before significant development occurred. The Everglades originally extended from north of Orlando all the way to Florida Bay on the south. The area was a swamp, hydrated by an annual average of sixty inches of rainfall per year, with two-thirds of that rainfall occurring in the summer wet season from May through October. Lake Okeechobee is often characterized as the "liquid heart" of the Everglades, storing water during the dry season and overflowing to create the "river of grass" during the wet season. The region was only accessible by boat until the late 1800s when Henry Flagler, whose wealth came from Standard Oil, constructed the Florida East Coast Railroad the entire length of the state, from Jacksonville to Key West, at that time Florida's largest city. Then people could come, and they did, initially as tourists and then as residents.

Because of the importance of water management to the environmental and economic well-being of the state, Florida is divided under state law into five water management districts with responsibility for water allocation (in Florida all water is owned by the state) to meet public and private beneficial purposes, flood-control purposes, and environmental protection related to water quantity and water quality. Southeastern Florida is within the jurisdiction of the South Florida Water Management District (SF-WMD), which covers sixteen counties and includes all of the historical Everglades system. The original mission of the SFWMD was flood control, and most particularly the operation of the Central and Southern Florida Flood Control Project, which was authorized as a federal flood-control project in 1948 and constructed by the U.S. Army Corps of Engineers thereafter. With more than 1,400 miles of canals and a series of levies and pumping stations, this is one of the most complex water management systems anywhere in the world. Figure 7-1 shows the network of canals, levees , and water conservation areas that blanket southeastern Florida from the southern tip of Lake Okeechobee at Belle Glade to the southern terminus of the Florida mainland south of Homestead. Counties and drainage districts operate secondary canal systems which convey street and property drainage to the regional system. How the system is operated determines where water is available for water-supply purposes, what land is flooded after major rain events, and where stormwater has an impact on natural systems such as Everglades National Park and Biscayne National Park. While the SFWMD is not the only regulatory agency overseeing water resources, it is the central state player in terms of water-supply and flood-control issues.

Southeastern Florida is extremely flat, ranging in elevation from about twenty feet above sea level in a few areas to sea level itself. Most of the area overlies the Biscayne Aquifer, an extraordinarily porous limestone wedge-shaped formation that is about 200 feet thick on the eastern shoreline and tapers out to nothing in the Everglades. Figure 7-2 shows the "Swiss cheese" quality of the limestone, which enables groundwater to move through it at a rate of up to ten feet per day under ambient conditions and much faster under the influence of a pumping well. This is one of the most productive aquifers in the world, but is also vulnerable to interaction with adjacent surface water and contamination from a variety of land uses.

Freshwater levels are highest to the west, driving the groundwater in a southeasterly direction, where it discharges into Biscayne Bay and Florida Bay to the south. Historically, groundwater levels were four to five feet higher than they are today, and in many areas the water was above the land surface during the summer wet season. The construction of drainage

FIGURE 7-1

South Florida Regional Water Management System

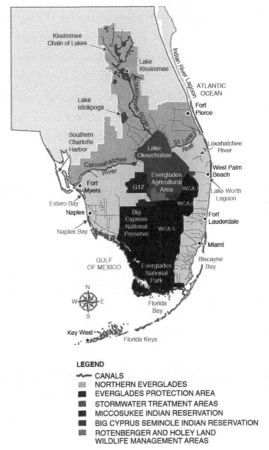

Source: South Florida Water Management District.

canals generally following natural drainage channels running west to east dropped groundwater levels significantly in the central and eastern portions of the region, keeping the land dry and useful initially for farming and then for extensive urban development. The flood-control system built pursuant to the 1948 federal authorization was designed to limit protracted flooding from rainfall typical of a storm event that was likely to occur once every ten years and to limit the duration of flooding that

FIGURE 7-2

Biscayne Aquifer Limestone

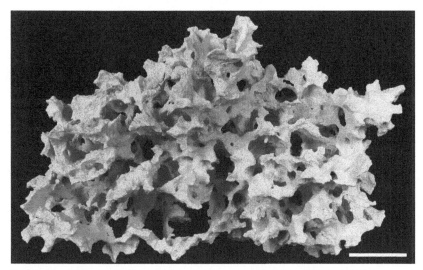

Source: Miami-Dade Water and Sewer Department.

would occur with less frequent but higher-intensity storms such as hurricanes. While this system does utilize pump stations in various locations to move water, most of the water moves by gravity from areas of high water to the west to area of lower water (sea level) to the east. By far the greatest volume of water that moves through the system does so by gravity, an issue of serious concern as sea level rises. When originally constructed, the drainage canals were uncontrolled at the coastline: so as long as freshwater elevations were higher than sea level, freshwater discharges would occur. During the dry season, incoming tides began the process of saltwater intrusion into the aquifer system, and structures had to be constructed to control excessive discharges of freshwater to limit saltwater intrusion during the dry season. Maintaining a balance throughout the system that limits flooding of urban and farm areas, maintains clean water to support the remaining Everglades system and coastal wetlands, and protects against saltwater intrusion into the region's primary source of drinking water continues to be the golden mean of water management operations.

Water Supply and Wastewater Management

Early settlers had no difficulty finding abundant, clean water. Individual wells could be very shallow, no more than fifteen or twenty feet. Sanitary wastes were channeled to the ground in slab-covered pits, and later through septic tanks and drain fields. As cities began to develop, mostly on the coast, sanitary sewers discharging into canals, bays, and the ocean were constructed, and water-supply utilities to meet water-supply needs more efficiently came into being, mostly operated by cities and counties. Originally both water-supply well fields and treatment plants and wastewater treatment plants were located near the coast to serve developed areas and to be accessible to receiving waters for disposal of sewage. As development began to move west and some well fields became salt-intruded, new water-supply facilities were located farther from the coast. In 1972 Miami-Dade County merged its utility systems with those of the City of Miami to create a countywide system, in part to respond to the new Clean Water Act opportunities for federal support to upgrade wastewater systems and to provide for the continuing migration of residents to Miami's "gold coast." One outcome of that decision was the construction of three large sewage treatment plants on the shoreline, two discharging treated wastewater into the ocean well offshore and one discharging effluent into deep injection wells. In addition, large water-supply well fields and treatment plants were constructed further west in locations designed to be safe from salt intrusion as the salt front was stabilized through the operation of the water management system. This approach was typical of the region, although the islands making up most of Monroe County (the Florida Keys) get most of their water from the mainland in Miami-Dade County because of the unavoidable saltiness of the island water.

Currently the southeastern Florida region has approximately fifty retail water-supply utilities and forty sewer utilities. Miami-Dade County has the most regional and interconnected system and by far the largest water-supply and sewage treatment system, producing about 300 million gallons per day (mgd) of drinking water and treating about an equal volume of wastewater. Total utility water-supply production in southeast Florida is about 500 mgd, most of which comes from the shallow Biscayne

Aquifer through water-supply wells that may be forty feet to eighty feet in depth, with the ability in some cases to produce up to 10 mgd from a single well. The secondary source of water that is beginning to be utilized in response to limits on additional withdrawals from the Biscayne Aquifer and the Everglades system is the Floridan Aquifer. Wells approximately 1,000 feet deep are needed to tap this brackish source of water. Reverse osmosis is the current treatment of choice for this water. For those reasons, the cost of water from the Floridan Aquifer is substantially higher than Biscayne Aquifer water. About 14 percent of water supplies for the lower east coast of Florida come from the Floridan Aquifer. One issue for the future is that the sustainable yield of the Floridan Aquifer has not been determined. Currently quantities in the range of 30 mgd are being produced either for drinking water or for power plant cooling water. Additional water-supply sources to meet future demands include wastewater reclamation and desalination of seawater. Southeastern Florida will never run out of drinking water, but it could certainly run out of inexpensive drinking water, which has always been part of its history.

Climate Change Issues Confronting Utilities in Southeastern Florida

Every region has its own set of climate change worries. In southeastern Florida those worries include storm surges affecting water and sewer infrastructure in coastal zones, the increasing incidence and duration of flooding, the increasing frequency or intensity of tropical storm events, a concentration of rainfall events that may increase both flood and drought events, progressive saltwater intrusion into the Biscayne Aquifer (requiring more expensive and energy-intensive treatment systems), rising groundwater levels, which could disrupt the functionality of thousands of septic tank drain fields in the region and substantially increase groundwater inflows into gravity sewer lines, and progressive failure of the regional flood-control system, which could result in substantial depopulation and a corresponding reduction in water and wastewater demands. Another important factor for southeastern Florida's future is the continued availability and affordability of property insurance. Continuous development has been a hallmark of the southeastern Florida economy, first characterized by single-family subdivisions spreading to farmland to the south and west and more recently by very rapid redevelopment of the historic coastal areas, bringing thousands of high-rise condominium and apartment units to downtown areas. Particularly since the temporary housing crash in 2008, the development market has relied increasingly on cash sales and foreign buyers. The rebound of construction and the housing market has been rapid, and the need to upgrade the old water and sewer infrastructure from, in some cases, its state a hundred years ago has become readily apparent. There is no question that development drives, and has always driven, the need for water and sewer infrastructure. On the flip side of that truism is the fact that future development and redevelopment cannot occur in the absence of water and sewer infrastructure. The difference today is the specter of climate change and sea-level rise, which could reverse the historical trend of continuous growth and even replace that trend with population contraction. Utilities, however, are not in a position unilaterally to determine that they will decline to meet tomorrow's needs in deference to what might happen the day after tomorrow. Only when there is a consensus in the private markets and the arena of governance that some risks in some areas are not practical to mitigate will there be a redirection of development to what may literally be "higher ground."

These climate issues are ripening at a time when much of the utility and water management infrastructure in the region is approaching asset life end and in need of serious and expensive attention. As an example, Miami-Dade County has identified a utility capital program of more than $13.5 billion that needs to be accomplished over the next fifteen to twenty years, reflecting in part the fact that much of its water and sewer infrastructure was constructed in the 1970s in conjunction with the Clean Water Act and the Safe Drinking Water Act. Grant funds, particularly from the Clean Water Act, encouraged a huge program of new assets, and now those assets are approaching age fifty. The same can be said for much of the regional flood-control system. While carrying a large price tag, this confluence of changing future conditions and the need for significant utility asset upgrade and replacement creates an opportunity to characterize and mitigate future risks that are likely to accompany climate change.

Organizing for Change on a Regional Basis

Awareness of and reaction to climate change began in southeastern Florida when Miami-Dade County became a participant with a dozen North American and European local governments in a greenhouse gas emission reduction project initiated by the International Council for Local Environmental Initiatives (ICLEI) in 1990. The purpose of the project was to develop a template that local governments around the world could use to assess greenhouse gas emissions within their jurisdictions and to formulate policies by which local governments, through their operations and in conjunction with other levels of government, could reduce emissions. A plan was adopted by the Miami-Dade County Commission in 1993, and since that time several advisory committees sequentially have addressed both greenhouse gas emissions and climate change adaptation across a number of jurisdictions in the region. As awareness of the likelihood of sea-level rise and its potential consequences increased, the need for coordination across local jurisdictions became apparent. In 2010, following a regional climate change forum in Miami, the Southeast Florida Climate Change Compact was created by joint action of Palm Beach County, Broward County, Miami-Dade County, and Monroe County. As noted on its website, the purpose of the Climate Change Compact is to coordinate mitigation and action plans across county lines and to coordinate jointly with other levels of government and stakeholders to encourage consistent approaches to planning and implementing climate change action plans.[2] The compact is addressing a variety of issues, but water management cuts across virtually all issues in terms of public water supply, flood control, wastewater management, sustainability of the region's natural features, such as the Everglades system and beaches, and the continuing viability of existing and future development patterns. The SFWMD, the state entity responsible for operating the Central and South Florida Flood Control project and for authorizing all water uses, including public water supplies, for a sixteen-county area that includes the Everglades system and the four counties composing southeastern Florida, is an active partner with the compact in undertaking the technical studies needed to support policy decisions.

FIGURE 7-3
Miami Beach Tidal Flooding

Source: Miami-Dade Water and Sewer Department.

Public and stakeholder engagement in the issue of climate change has been enhanced substantially by the annual conferences sponsored by the Climate Change Compact. That the organization was created by elected officials from each county who took on the responsibility of being champions for climate change policy as a regional issue is significant. Public awareness of the issue has been elevated by a series of coastal flooding events that were not associated with hurricanes and in some instances were associated only with extremely high tides ("king tides") on clear days that caused street flooding when ocean water backed up through drainage systems into the street. Figure 7-3 is a photograph of such an event on Miami Beach, the city that has been most reactive to vulnerability. The city is actively implementing a program to construct more than sixty stormwater pumping stations and to elevate selected streets as a hedge against the increasing frequency of street flooding resulting from sea-level rise. This will represent an investment of more than $400 million.

A direct outcome of regional collaboration with respect to the integrity of the water supply has been agreement on a common approach to modeling the potential for saltwater intrusion and corresponding implications for the future water supply. Miami-Dade and Broward Counties have contracted with the U.S. Geological Survey (USGS) to develop interactive surface water–groundwater models, and this work is being reviewed and coordinated with the SFWMD modeling group. The Climate Change Compact has adopted the U.S. Corps of Engineers (USCOE) projections for sea-level rise, a bracketed projection that assumes at the low end of the projection the implementation of some measures to limit future greenhouse gas emissions and at the high end the lack of effective restrictions on greenhouse gas emissions. An extension of the USCOE high-end projection to 2075 is an increase of sea level by about three feet. Using the same scale, the high-end projection for 2030 is expected to be about seven inches. The porosity of the Biscayne Aquifer providing the shallow groundwater is extraordinarily high, and the limestone interacts directly with the saltwater from Biscayne Bay and the Atlantic Ocean. The only effective barrier to saltwater intrusion is to maintain a sufficient elevation of freshwater. Dikes and bulkheads built on top of the ground will not prevent saltwater from moving inland as the sea level rises.

The regional water management system has been operated over the past sixty years to stabilize saltwater intrusion near the coast by keeping groundwater levels high, but not so high that land flooding occurs. This has been a largely successful strategy, and the salt barrier line has been remarkably stable. Areas of greatest risk in terms of public water supplies have been well fields located relatively close to the coast, such as in the city of Hollywood in Broward County. During the dry winter season, when public water-supply demands tend to be highest, production from these coastal well fields has been limited, especially in drought periods. Moving well fields to the west mitigates the immediate risk of salt intrusion, but over the longer term sea-level rise can create vulnerability even of more westerly locations. Initial model analysis of Miami-Dade County using the USCOE sea-level-rise projections confirms there is likely to be little movement in the salt barrier line out to 2030, assuming that the regional water management system continues to be effective. Figure 7-4 schematically

FIGURE 7-4

Salt Intrusion Line

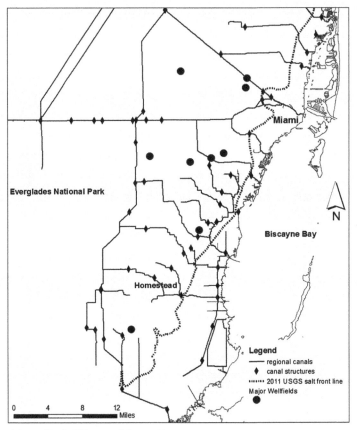

Source: Miami-Dade Water and Sewer Department.

depicts the salt barrier line in Miami-Dade County as it is now and as it is forecasted to be under 2030 conditions. It suggests that Miami-Dade County can continue to rely on current water-supply well fields and treatment systems to provide drinking water to area residents at least until that time. Additional work is under way to forecast over a longer time frame and to incorporate regional water management system modeling outcomes to further predict salt intrusion probabilities in the region beyond 2030. The current technology of choice to remove salt content from source water is membrane filtration and reverse osmosis. This treatment is currently being used to treat water from the brackish Floridan Aquifer in a few lo-cations in southeastern Florida, and it is likely to be increasingly the water-supply technology of the future, as least until a less energy-intensive and less costly technology is developed.

Storm Surge and Climate Change: How Much Mitigation Is Affordable?

The history of southeastern Florida is in significant part the history of exposure to tropical storms by virtue of its location in the subtropics. The great hurricane of 1926 that roared through downtown Miami effectively ended the development boom that followed the arrival of Henry Flagler's Florida East Coast Railroad in 1898. In 1935 another powerful storm hit the Florida Keys, destroying parts of the railway that crossed seven miles of open water on its way to Key West. Significant development did not resume until after the Great Depression and after World War II, when thousands of returning GI's and their families began to come to Florida for the weather and the opportunities of a growing community. Vacation-ers and retirees came for both sun and the glamour, highlighted by fre-quent visits of the rich and famous. Tropical storms and hurricanes happened from time to time, but until Hurricane Andrew in 1992, the most expensive natural disaster in history at the time it occurred, the risk of tropical storm activity was viewed as manageable and not at all out of line with the earthquakes, floods, tornadoes, and fires that occur in other parts of the country. The approach to mitigating risk to tropical storm events focused on three strategies: (1) the elevation of structures to limit flood damage from storm surge and the limitations of the drainage sys-tem; (2)a building code requiring that structures be able to withstand hur-ricane force winds; (3) and an emergency planning and response system to evacuate residents and visitors from the most vulnerable locations (such as barrier islands, all of the Florida Keys, and coastal locations) in advance of storms and providing public shelters in schools and other facilities. De-spite the enormous damage to property that occurred from Hurricane Andrew, only a handful of deaths were attributed to the storm, perhaps in part because it was a compact storm that went through the less-developed southern portion of Miami-Dade County, but also in part because of the mitigation of risk. What did occur as a result of the experience (in addition to a further strengthening of the building code) was a dramatic increase in the cost of insurance. Climate change introduces new factors into the risk equation. As temperatures warm, the frequency and intensity of tropical storm activity may increase. Historically there have been more and less active periods for tropical storm activity, sometimes spanning several decades. A decided increase in the frequency of storms can have both physical and psychological consequences. Sea-level rise will add to the consequences of storm surge, potentially creating greater destructive force, beach erosion, and flooding in the areas where redevelopment has been the most intense over the past fifteen years.

Water and sewer utilities are vulnerable to storm-surge risks at critical facilities located in coastal storm-surge zones. Treatment plants have always been at risk of direct damage and power loss from tropical storms. Most such facilities have emergency power capabilities (the entire Miami-Dade system can operate entirely off the grid). Historically, actual damage has been infrequent and manageable, but that experience may not be a reliable indicator of the future. Hurricane Andrew tracked quite close to Miami-Dade's South District Wastewater Treatment Plant, bringing a sixteen-foot storm surge through the area. The backup generators were flooded and inoperable after power was lost from the grid. Some other damage occurred, but the plant could operate after power was restored about two weeks after the storm. As it turned out, the extensive property damage to homes and buildings served by the plant caused most people to evacuate, and the sewer plant being out of service for two weeks as a result of the loss of electrical power was not a consequential factor in terms of public health or the recovery effort. This experience illustrates that damage from hurricanes at treatment works can be repaired before service demands in the area served by those treatment works return to pre-storm conditions. This may be different, in terms of risk assessment and mitigation, from events such as tornadoes and earthquakes, which can be relatively localized and could selectively hit a treatment plant without simultaneously damaging significant portions of development in the service area of the plant, thereby creating a water-supply deficit with public health implications.

There are no specific standards by which utilities assess and mitigate storm-surge risk. This became evident recently when interveners in a Clean Water Act consent decree negotiation regarding Miami-Dade County's three coastal sewer plants recommended that the resulting consent decree explicitly acknowledge and incorporate requirements to mitigate additional risk to the plants resulting from sea-level rise. The consent decree approved by the federal court did not include any such standards or requirements, although it was clear in the record that the county was committed to incorporating a risk analysis and a range of mitigation measures into the substantial reconstruction of the plants that is required by the consent decree. The approach to assessing risk involves a comparison of the existing elevation of various plant components with flood levels that are likely to occur in the future under ambient (non-storm) conditions and under various hurricane conditions, augmented by an accepted forecast of sea-level rise.

An initial estimate was made using actual storm-surge data from Hurricane Andrew and adding a three-foot sea-level-rise factor (the USCOE's high-end forecast for 2075), plus extreme high tide conditions. This analysis, including interpolations for category 1 through category 5 hurricanes, suggests that plant components that historically would be flooded under category 3 hurricane storm-surge conditions would in the future be at a similar flood risk from a category 2 hurricane. Ambient conditions at the plants with a three-foot rise in groundwater levels (which will rise with sea level) would not result in constant flooding at the plant sites (though it could cause constant flooding in areas served by the plants). Figure 7-5 illustrates this analysis for Miami-Dade County's Central District Wastewater Treatment Plant, the largest sewer plant in Florida. It is located on a barrier island in Biscayne Bay, so it is also highly vulnerable to storm surge. The figure shows plant facility elevations, varying from ten feet to eighteen feet above sea level, compared with predicted storm surges, including sea-level rise out to 2075, for various categories of hurricanes.

The preliminary study identified electrical components and control systems to be at greatest risk, with hard structures such as tankage being at less risk. Risk-mitigation options include relocating the plants altogether (very expensive), constructing barriers around the entire plant facility to repel and exclude the storm surge, and flood proofing individual components

FIGURE 7-5

Facility Storm Surge Analysis

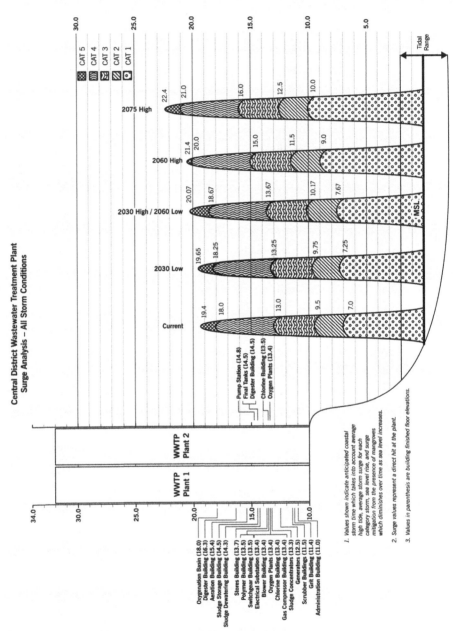

Central District Wastewater Treatment Plant
Surge Analysis – All Storm Conditions

1. Values shown indicate anticipated coastal storm time which takes into account average high tide, average storm surge for each category storm, sea level rise, and surge mitigation from the presence of mangroves which diminishes over time as sea level increases.

2. Surge values represent a direct hit at the plant.

3. Values in parenthesis are building finished floor elevations.

Source: Miami-Dade Water and Sewer Department

of the plant either by elevating components (such as emergency generators and control systems) or by hardening the facilities to resist the entry
of water under storm-surge conditions (the most cost-effective approach).
The ability to maintain hydraulic flow through the plants was determined
to be the most critical factor in terms of mitigating the short term risk
from storm surge. The site-specific characteristics of each plant must be
considered in the determination of a cost-effective plan for managing the
storm-surge risk at each location. The probable asset life of each plant
component will be a factor in determining whether a particular mitigation approach should be applied now or whether that investment should
be delayed to a future asset replacement cycle when sea-level rise and
tropical storm projections may be more certain.

Water-Supply Considerations

Water treatment infrastructure, while typically not at risk from storm-
surge issues, is subject to flooding and windstorm damage, along with all
other aboveground structures. This means that risk assessments for flood
conditions based on higher groundwater levels and a potential decline in
the efficiency of drainage systems should be undertaken in the region as
water management modeling incorporating projected groundwater levels
is completed and flood mitigation approaches are evaluated. Already in
place in Miami-Dade County is a directive that all county infrastructure
projects are to include a fifty-year (or life of asset, if that is longer) analysis
of sea-level-rise impacts so that risk mitigation can be incorporated into
the design of the facilities. Structural windstorm risks have long been addressed in Florida's building code, but additional reviews are likely to occur in light of the actual experience with future hurricanes, the evolution
of building materials and techniques, and improvements in understanding the relationship between climate change and the frequency, intensity,
and tracking of tropical storms. In addition to water treatment plants,
distribution facilities such as storage tanks and repumping stations (which
could be at risk of storm-surge impacts in some cases) need to be considered in terms of risk assessment and mitigation.

The need for ongoing assessment of salt intrusion into the freshwater supply has already been described. Most utilities in southeastern Florida operate on a twenty-year water-use permit from the state, with regional water-supply plans updated every five years by the SFWMD. Future demand projections can vary greatly based on future population projections and per capita consumption trends. The current regional water-supply plan is now forecasting about a 10 percent reduction in future demands compared with the previous regional water-supply plan. Utilities are obligated to demonstrate how they will meet their future supply needs, with potential development constraints if required infrastructure is not constructed in time to meet the demands. As the consequences of climate change and sea-level rise become more defined in the future, the importance of these demand forecasts, which could at some point include population declines in lieu of the historical population growth, will increase for utilities as they try to match infrastructure investments with the customer base that will utilize and pay for those investments in the future.

Meeting the Climate Change Challenge in Southeastern Florida

Utilities everywhere are dealing with aging infrastructure, limited financial resources, an increasingly complex regulatory environment, new technology opportunities, and the additional consideration of climate change consequences that may have an impact on supply sources, infrastructure integrity, treatment requirements, and future demands. In southeastern Florida, all of these factors are in motion on a regional basis. While utilities will experience climate change consequences in different ways according to their specific circumstances, there are some initial conclusions and lessons already learned as the certainty and nature of change become more apparent and public awareness and expectations are elevated.

1. Regional collaboration among local government jurisdictions and among utilities will be valuable in terms of standardizing forecasts, economizing on and jointly benefiting from technical analyses, attracting technical assistance and funding from state and

federal funders and foundations, and maintaining stakeholder focus on the implications of climate change over time. Collaboration also underscores the need for attention across a number of infrastructure and service areas typically not under utility control. Most prominent in southeastern Florida is the long-term effectiveness of the regional and local drainage systems, the control of development patterns, and the continued availability and affordability of flood and windstorm insurance.

2. As infrastructure is renewed, replaced, or added, opportunities to mitigate climate change risks need to be assessed and incorporated into design and construction, taking into consideration probable asset life. This practice needs to become an element of every utility's planning process.

3. While adaptation to consequences is now a primary focus of climate change planning in southeastern Florida, energy and carbon efficiency should not be neglected. In the long term, much of Florida's future viability will be a function of the extent to which atmospheric carbon concentrations can be stabilized, as will be the case for many coastal communities around the world.

4. Technology research and innovation may be critical in mitigating various climate change risks, including water treatment and reclamation, distributed versus centralized systems, flood proofing and storm-surge protection techniques, and modeling and forecasting tools. Support for and participation in the research enterprise by utilities, universities, private sector suppliers, and all levels of government require continued focus going forward. Part of this research agenda will document the successes (and failures) of risk-mitigation projects around the world. Utility managers need to be positioned to learn and benefit from these experiences.

5. Adaptive management that incorporates progressively more reliable sea-level-rise forecasts with near- and medium-term demand forecasts as the basis for infrastructure investment is the basis on which these investment decisions must be made. Utilities are

obligated to meet near-term demands, even if long-term demands may contract. Overinvestment in mitigation measures addressing long-term forecasts can result in excess capacity if future demands do not materialize or climate change consequences are less severe than anticipated, while time may bring less expensive solutions through research and development. Underinvestment, of course, carries its own set of risks. Utilities may make all the "right" decisions and investments, but if other service providers, both public and private, do not, there will be consequences for utilities.

6. Effective engagement of stakeholders, including civic, business, environmental, and customer groups, will continue to be an essential aspect of an adaptive management program responding over a long time horizon to climate change issues. This engagement sharpens considerably when the discussion involves actual experience in the community of events that are recognized as being associated with climate change. In southeastern Florida this experience has been focused on periodic drainage system failures in coastal communities when clear-sky street flooding has occurred at high tides. Other types of indicators may include salt intrusion into coastal water-supply wells, a measurable elevation of ambient seasonal groundwater levels, and the availability and cost of property insurance. This is the point at which planning can most effectively move to action, with the investment necessary to support that action. Utility managers need to be ready when opportunities arise.

The next twenty years of experience are likely to sharpen considerably the adaptive management strategies of utilities in southeastern Florida, providing more direct evidence of the rate of change in sea-level rise and more certain predictions of tropical storm tendencies. Some risk-mitigation measures now on the drawing board will be able to benefit from past experience, and research may yield new and better future options. Regional plans helping to mitigate nonutility risks relative to drainage and insurance may be in place, thereby better defining future development options and utility demand forecasts. What utility managers in southeastern Florida cannot afford is disengagement from the challenges and uncertainties that are now clearly established in the region. They would do so at the peril of their utilities and their customers.

Notes

1. Nicholls, R. J., et al. (2008), "Ranking Port Cities with High Exposure and Vulnerability to Climate Extremes: Exposure Estimates," OECD Environment Working Papers, No. 1 (Paris: OECD Publishing). DOI: http://dx.doi.org/10.1787/011766488208.

2. Southeast Florida Climate Compact (http://www.southeastfloridaclimatecompact.org).

Chapter 8

Finding the Balance in New York City
Developing Resilient, Sustainable Water and Wastewater Systems

ANGELA LICATA AND ALAN COHN

New York City's water network is one of the largest systems in the world (see figure 8-1). Through a complex arrangement of dams, reservoirs, and aqueducts, the New York City water-supply system has a capacity of 580 billion gallons. It serves 8.4 million New York City residents, millions of daily commuters, and more than a million residents in communities north of New York City. After traveling up to 125 miles from the New York City water-supply watershed, water is distributed by more than 7,000 miles of water mains and pipes relying almost entirely on the pressure from gravity to deliver water up to six stories above street level. Water that enters the city's drains is conveyed through 7,500 miles of sewers and returned to New York City waterways.

Sustainability and long-range planning are fundamental to New York City's water-supply, drainage, and wastewater systems. More than 150 years before New York City's first sustainability plan was formulated, the water supply was developed, and then expanded over the course of a century, to protect the public from fire, support public health, and accommodate growing populations. Water-supply sources were located in pristine areas outside the city and water was conveyed by the force of gravity. The city's

FIGURE 8-1

New York City's Water Supply System

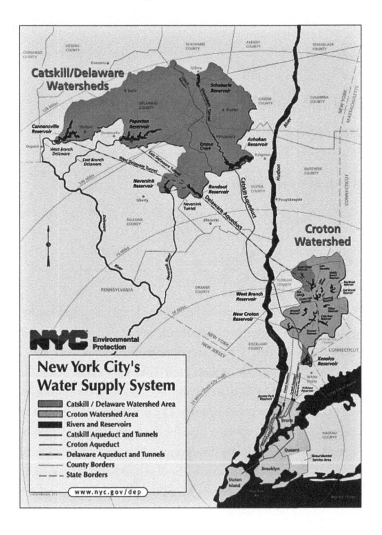

first aqueduct, delivering water from the Croton Watershed in Westchester County, was completed in 1842. Around that time the city began to recognize the need to clean its harbor and coastal waterways to maintain good water quality for fishing and recreation. As a result, the city built the first wastewater treatment plant at Coney Island in 1886. As the city grew, so did the water and wastewater systems, with new reservoirs drawing from more remote locations at higher elevations and new treatment plants strategically located near beaches and fisheries, at low-lying points to receive flow mostly by gravity.

More recently, New York City recognized that the system already had the foundation for a hybrid approach that incorporated engineering and natural solutions to preserve water quality in reservoirs. Starting in 1993 and continuing today, New York City is one of only five large municipalities that receive a waiver from the U.S. Environmental Protection Agency (EPA) filtration requirements under the Safe Drinking Water Act (SDWA). The waiver recognizes the capacity of New York City's system to naturally treat its high-quality waters, thus avoiding the construction—and associated energy and maintenance requirements— of a multi-billion-dollar treatment facility. Nonetheless, regulatory requirements and the need to manage water quality in a mostly unfiltered system have required engineering approaches. This has resulted in an increase in energy use and associated greenhouse gas emissions. However, the gravity-fed system limits energy demand, thus mitigating the emissions that would have been associated with pumping. Furthermore, the size and the design of the extensive gravity-fed system of reservoirs have created opportunities to harness hydropower, which already generates almost 180,000 megawatt-hours of electricity each year.

Similar innovations have emerged more recently for managing stormwater and wastewater in the city. Beginning in the 1990s, the city's Department of Environmental Protection (DEP) pioneered the Staten Island Bluebelt system, restoring wetlands and streams for storage and conveyance as an alternative to raising streets and constructing more sewers. More recently, in 2010 DEP initiated its Green Infrastructure Program to capture stormwater at the source, thereby reducing the discharges and overflows of untreated water from combined sewer systems, and thus reducing the city's reliance on large tanks and tunnels and the associated costs of construction, operations, and maintenance.

Amid these successes and innovations, new challenges have emerged with recent climate extremes. While it is impossible to attribute any one event entirely to climate change, extreme events are evident in the sea-level, precipitation, and temperature record, with clear trends toward increasing volumes and variability. In 2011, Tropical Storm Irene dumped up to sixteen inches of rain in some parts of the New York City watershed, flushing large volumes of sediment into reservoirs. Just ten days later, Tropical Storm Lee produced more heavy rain before the system could recover from Irene, prompting unprecedented levels of reservoir turbidity and coliform bacteria. Storms like Irene and Lee contribute to annual precipitation amounts that have increased at a rate of approximately 0.8 inches per decade since 1900, with large year-to-year deviations from the average especially since the 1970s.[1] Higher sea levels also increase the extent and magnitude of coastal flooding caused by storms. In 2012, Hurricane Sandy inundated New York City with its record storm surge, flooding many waterfront communities and causing sewers to back up into homes and untreated sewage to spill from damaged wastewater treatment plants. Since 1900, sea levels have risen more than a foot in New York City, primarily as a result of climate change. As sea levels continue to rise, coastal storms will cause flooding over a larger area and at increased heights than they otherwise would have.

Today the stakes are higher, with more people and property at risk owing to the changes in climate that have already occurred, and with regulatory requirements that continue to increase under the Clean Water Act and the SDWA. Water-supply systems must often project capital planning for thirty to fifty years in advance to meet the competing demands of ensuring an adequate, good-quality supply in future while maintaining the infrastructure in a state of good repair and keeping water rates affordable and equitable. Addressing climate change, including the impacts of extreme events superimposed on increasing temperature, precipitation, and sea level, adds an important new dimension to the long-range planning process. These competing demands have required DEP to enhance its efforts to address the long-range sustainability, resiliency, and adaptability of the water and wastewater systems. Taking on these challenges requires planning foresight, engineering innovation, and cooperation with environmental regulators, customers, and the public.

Drinking Water Supply and Distribution

Water supply for DEP customers is collected and transported from a network of nineteen reservoirs and three controlled lakes in a 1,972-square-mile watershed that extends 125 miles north and west of the city; the total capacity of the system is 580 billion gallons. In addition to the visible aspects of the city's water supply (reservoirs, dams, and waterworks), DEP operates and maintains three major subsurface aqueducts and their associated shafts, culverts, and siphons: the twenty-four-mile long New Croton Aqueduct, the ninety-two-mile long Catskill Aqueduct, and the eighty-five-mile long Delaware Aqueduct. Both the Catskill and Delaware Aqueducts cross under the Hudson River. Other tunnels in the system convey water between reservoirs and the distribution system. In addition, pumping stations at Croton Falls and Cross River reservoirs provide the ability to transfer water from the Croton system to the Delaware Aqueduct, which is critical to the operational flexibility of the system.

The numerous components of the city system combine to create a robust and flexible supply network. Under normal conditions, the Catskill and Delaware systems west of the Hudson River contribute 90 percent of the city's surface water supply, while the Croton system east of the Hudson provides the remaining 10 percent. At times the Croton system supply can be increased to up to 30 percent of the total. For several years, however, the Croton system was off-line during construction of the Croton Water Filtration Plant, and the city drew 100 percent of its water from the Catskill and Delaware systems. When the Croton Water Filtration Plant became operational in 2015, the Croton system water was again made available, restoring the water supply system to its full capacity.

In addition to operational flexibilities, naturally occurring factors also contribute to the reliability of the water supply. These natural foundations also support high water quality: the Catskill and Delaware Watersheds are 75 percent forested, the system is almost exclusively gravity-fed, and it benefits from a historically temperate climate. These elements help safeguard the system against service interruptions by controlling water quality issues linked to turbidity, pathogens, and other contaminants that could affect millions of people and hamper provision of some of the city's essential services.

DEP manages, operates, and protects this vast public water-supply system to meet multiple objectives, primarily to provide high-quality potable water to the residents of New York City. Throughout the onset of environmental regulations like the SDWA, the city provided potable water that met stringent drinking water quality standards. The quality of the water supplied has met the requirements of ever-expanding regulations. More recently, city operations have been recognized by federal and state regulators as qualifying for a hard-to-achieve exemption provided in the Surface Water Treatment Rule (SWTR). That rule requires surface water purveyors to filter potable water before distribution to consumers. Since the early 1990s, federal and state regulators have issued a Filtration Avoidance Determination (FAD) for the Catskill and Delaware systems based on the high quality of the city's source waters as provided for under the SWTR and the accompanying treatment and watershed management methods, extensive monitoring, and effectiveness of the source water protection program. Source water protection is a particularly important aspect of the FAD and includes enforcement of watershed rules and regulations, the operation of city-owned wastewater treatment plants in the watershed at high levels of treatment, upgrading of all noncity wastewater treatment plants in the watershed, long-term protection of lands through acquisition in fee or easement, and partnership programs to support watershed landowners and communities in protecting source water quality. Maintaining the FAD not only affirms the city's top-level service but also avoids the need to build a filtration facility for 90 percent of the city supply, which would undoubtedly cost the city billions of dollars in construction and operations.

Other water-supply objectives include sanitation, firefighting, and commercial and recreational purposes. Additionally, DEP diverts or releases water from the system to supply communities outside the city, maintain flow in the Delaware River, support fishery habitats and recreation, generate electricity, and further enhance the flood mitigation provided by the dams to downstream areas. These secondary purposes are all governed by an ever-evolving collection of agreements, operating protocols, state and federal laws and regulations, court decisions, and consent decrees. Releases from Delaware system reservoirs to maintain flow in the Delaware River are governed by a 1954 Supreme Court decree and subsequent agreements among the decree parties, which allow the city to divert up to 800 million gallons per day (mgd) from the river's headwaters for its water supply. Balancing the needs and requirements for downstream releases with city water-supply demand is a critical function for DEP and necessitates cooperation and working relationships with numerous stakeholders, including states and other water utilities that draw from the Delaware River.

To continually supply top-quality water, DEP works to anticipate and address potentially recurring problems in the system through design elements. For example, DEP has implemented measures to address turbidity in the Catskill System streams, a naturally occurring phenomenon resulting from the geology of the region. When New York City's engineers designed the Catskill water-supply system, episodic turbidity was recognized as a water quality issue and ways to combat it were incorporated into infrastructure design. Simple approaches were then implemented to let turbidity from the Catskill's clay soils naturally settle from reservoirs before entering system intakes. For instance, the Ashokan dividing weir separates the reservoir to allow sediment time to settle in the west basin, before water enters the east basin, on its way to the Catskill Aqueduct.

Despite efforts to anticipate operational difficulties, extreme events and climate change may bring potential challenges to the long-range operability of the supply system and the city's continued ability to meet the criteria of the FAD and the SWTR. As a result, DEP has been devoting increasing resources to studying how even gradual shifts in climate might affect the system in the future. The Climate Change Integrated Modeling Project, or CCIMP, was initiated in 2008 to evaluate the effects of future climate change on the quantity and quality of water in the New York City water supply.[2] The CCIMP is designed to address three issues of concern to New York City: (1) the overall quantity of water in the entire water supply, (2) turbidity in the Catskill system and Kensico reservoirs, and (3) eutrophication in the Delaware system reservoirs.

In the first phase of the project, an initial estimate of climate change impacts was made using available global climate model data sets and DEP's suite of watershed, reservoir, and system operation models. Initial results from the CCIMP suggest that streamflow would increase during the late fall and winter and decrease in spring, with a shift toward more rain and less snow during the winter, as well as earlier melting of the typically smaller snowpack that would develop. Greater winter streamflow could cause reservoirs to fill earlier in the year and spill from the reservoirs to increase during the winter. The shifting seasonal pattern in streamflow could also result in increased turbidity in the autumn and winter but decreased turbidity in the spring.

As an unfiltered system, the potential effects of climate change and changes in the frequency and severity of extreme weather events are an important concern for the city. Under the SWTR, turbidity cannot exceed 5 nephelometric turbidity units (NTUs) at a point just prior to primary disinfection because of concerns about the effect of turbidity on disinfection effectiveness. When turbidity levels are too high, treatment with aluminum sulfate (alum) can be used to coagulate and settle particulate and colloidal matter in the system that might otherwise interfere with disinfection. The combination of Tropical Storms Irene and Lee in 2011, for instance, led to a record 260-day alum treatment regimen for the Catskill system. However, New York State regulators require costly and operationally disruptive dredging of alum flocculate. DEP is therefore employing a "multibarrier" approach to managing turbidity to meet water quality standards and ensure that communities tapping into the city's system upstate have sufficient water supply to meet their needs.

As part of the multibarrier approach, DEP is using cutting-edge technology to make the best operational decisions in order to minimize turbidity incidents. The Operations Support Tool (OST) is DEP's state-of-the-art decision support system that integrates multiple sources of critical near real-time operations data—streamflow data, in-reservoir water quality data, SCADA data, and current infrastructure information—into an advanced version of DEP's existing water-supply/water quality model. OST leverages DEP's monitoring and modeling investment under the FAD to understand the physical, chemical, and biological processes in the watershed affecting water quality and the substantive effort to model terrestrial, hydrodynamic, and eutrophication processes. OST taps into a network of data from water quality, hydrological, limnological, and meteorological sensors that provide input to the model. OST combines current system data with inflow forecast data and system operating rules to make probabilistic projections of reservoir levels and water quality over the coming weeks and months. This look-ahead capability provides system analysts, operators, and managers with information to support decisions concerning reservoir diversions and releases, and allows operators to test the risk and reliability of actual operations decisions on paper before implementing them.

Another element of the multibarrier approach to managing turbidity is operation of a release channel at the Ashokan Reservoir to discharge turbid waters into the Lower Esopus Creek prior to or during storm events to provide storage for capturing and settling. The creek's status as an impaired water body under section 303(d) of the Clean Water Act results in competing regulatory requirements. Therefore, the operation of the release channel is currently under environmental review in accordance with a state order on consent. To be a viable option for managing turbidity, the operating protocol for the release channel must take into consideration competing priorities of water quality (SDWA/SWTR goals for drinking water, Clean Water Act goals for the Lower Esopus, and minimizing chemical treatment under the State Pollutant Discharge Elimination System), enhanced flood mitigation for downstream communities, and ecological needs for the creek.

DEP is taking action to ensure a sufficient supply of high-quality water even when a turbidity event is unavoidable. A new connection between the Catskill and Delaware Aqueducts, known as the Shaft 4 Connection, will allow DEP to divert Delaware system water into the Catskill Aqueduct. This will occur at the existing Shaft 4 Connection site located at the point where the two aqueducts cross. This will allow DEP to reduce the flow of water from Ashokan Reservoir when turbidity is elevated while still maintaining sufficient flow to provide service to outside communities and meet overall system water demand. This increases operational flexibility, reduces turbidity levels entering the Kensico Reservoir (by blending Catskill diversions with low-turbidity Delaware water), and improves water quality for outside communities.

Optimization and efficiency of the existing infrastructure are fundamental to the sustainability of the system, particularly in times of drought and other extreme weather events that affect supply. The city is also preparing for repairs to the Delaware Aqueduct, which conveys, on average, 50 percent of the city's water from upstate sources. This aqueduct has been leaking between 15 and 35 mgd for many years. DEP is constructing a three-mile bypass tunnel around the section that has the largest leak. While the bypass is connected and the aqueduct is out of service, DEP will repair other sections of the tunnel. Since the Delaware Aqueduct will need to be shut down to connect the new bypass tunnel, there will be a temporary decrease in available water supply. The tunnel shutdown, repairs, and reactivation are expected to be completed in 2022.

In preparation for the shutdown, DEP is increasing the capacity and use of the Catskill and Croton systems and adopting both a new Water Demand Management Plan to conserve water citywide and water shortage rules to impose use restrictions during droughts and infrastructure repairs. The Water Demand Management Plan targets a 5 percent overall reduction in water consumption citywide by 2020 through municipal, residential, and nonresidential water efficiency programs and water distribution system optimization. Although designed to meet the more immediate needs for the Delaware Aqueduct repairs, DEP's demand management strategies provide long-term benefits by reducing the overall throughput of water and therefore the energy used in the new Catskill Delaware Ultraviolet Disinfection and Croton Water Treatment Plant facilities, and also for pumping and pollution reduction in in-city wastewater treatment plants. In addition, it allows DEP to accommodate population growth and increase drought resiliency.

Wastewater Collection and Treatment

As the water supply system grew, so too did the city's sewer network. At the beginning of the twentieth century, sewers discharged untreated wastewater containing a wide array of disease-carrying microorganisms into the city's waterways. To address this public health issue, the city built its first wastewater treatment plant at Coney Island in 1886, followed by two more in Jamaica Bay in 1894 and 1903. These plants treated only a fraction of the city's total wastewater flow, but they did marginally improve the quality of the hugely popular beaches at Coney Island. With new funding available between 1945 and 1965, five new plants were built to meet the needs of the growing population of New York City, which was approaching 8 million.

Throughout the 1960s, the effects of pollution became increasingly apparent, and concern for the environment was on the rise around the nation. In New York City, pollution control efforts were reflected in continued investments in wastewater treatment. By 1968, twelve wastewater treatment plants were operating in New York City and were capable of treating 1.4 billion gallons of wastewater each day. From the late 1970s on, spurred by passage of the Clean Water Act, federal funding, and support from the growing environmental movement, New York City continued to upgrade and expand its wastewater treatment system. Finally, in 1987, the city completed construction of two new treatment plants, and with a total of fourteen wastewater treatment plants, essentially all of the city's dry-weather wastewater flows were captured and treated for the first time.

During this period, the city also began to address combined sewer overflows (CSOs). Like many older cities, New York has a combined sewer system, which means that the sewers accept a combination of both sanitary and stormwater flows. In dry weather, virtually all of New York City's sewage is treated. During rainfall, however, the added volume of stormwater can exceed the capacity of the combined sewer infrastructure. This can result in untreated discharges from relief structures that are designed to protect the biological treatment process in treatment plants and to prevent sewage backups and flooding. Treatment plants are designed to manage twice the volume of wastewater on a wet day than on a dry day, but changes in precipitation, population, and impervious cover have increased the flow to the plants and therefore the volume of combined sewer overflow. In an effort to capture and treat more sanitary sewage flow during storms and prevent CSOs, in 1972 the city began to operate its first combined sewer overflow facility at Spring Creek, an inlet of Jamaica

FIGURE 8-2

Bluebelt System Preserving Natural Drainage Corridors to Convey
Stormwater

Bay. Infrastructure upgrades have enabled the city to increase stormwater capture rates from 18 percent in 1987 to 78 percent today.

As the city finished building wastewater treatment plants, it also began to move away from building new combined sewer systems, so that
today almost 40 percent of the city is served by separate sanitary and
stormwater systems—thus avoiding combined sewer overflows in these locations. Furthermore, starting in the 1990s DEP began to preserve natural drainage corridors, called bluebelts, including streams, ponds, and other
wetland areas, to convey stormwater (see figure 8-2). The Bluebelt program,
which began and has expanded most significantly in the borough of Staten
Island, has saved tens of millions of dollars in infrastructure costs compared with providing conventional storm sewers to achieve the same level
of service for the same land area.

In addition to controlling stormwater through bluebelts, in the early
2000s the city committed to evaluate innovative stormwater management
strategies through an interagency Best Management Practices Task Force.
The Sustainable Stormwater Management Plan, released by that task
force in 2008, concluded that green infrastructure was feasible in many

FIGURE 8-3

Green Infrastructure "Bioswales" Intercepting Stormwater Runoff
before the Water Enters the Combined Sewer System

areas of the city and could be more cost-effective than certain large infrastructure projects such as CSO storage tunnels.[3] Based on these recommendations—and building on the success of other efforts to build ecological system in the upstate watershed, the Staten Island Bluebelt, and Jamaica Bay—in 2010 DEP released the NYC Green Infrastructure Plan.[4] The plan proposed to utilize green infrastructure to improve the quality of waterways around New York City by managing runoff from impervious surfaces in combined sewer watersheds through detention and infiltration source controls. Green infrastructure uses plants, permeable areas, and other source controls, in a decentralized and integrated network to mimic the predevelopment water cycle and reduce stormwater runoff (see figure 8-3). It can also provide other benefits that will make New York City more sustainable by making more effective use of the city's natural resources and reducing energy costs in the process.

The Green Infrastructure Plan was proposed as an adaptive management strategy—an iterative, flexible decisionmaking process whereby incremental measures are continually evaluated and improved. An adaptive management approach was considered essential because of the magnitude of investment required to manage stormwater and the wide range of uncertainties about future conditions, including changing climate, rainfall, population, water demand, land use, technology, and regulatory requirements. DEP modeling showed that the green strategy would reduce more CSO volumes at significantly less cost than the all-gray strategy that was then included in New York City's CSO Consent Order and Waterbody/Watershed Facility Plans, submitted to the New York State Department of Environmental Conservation (DEC). Moreover, the significant sustainability benefits of the green strategy, such as urban heat island reduction, mitigation of air pollutants and carbon dioxide, and increases in property value, would begin to accrue immediately and build over time, in contrast to the benefits of building tanks, tunnels, and expansions, which would yield noticeable water quality benefits only at the end of a decades-long design and construction period. Because of the sustainability benefits, in March 2012 the city and DEC finalized an agreement that incorporated the iterative, adaptive management approach to sustainable stormwater management using green infrastructure.

Innovative new systems such as the Bluebelt program and Green Infrastructure Plan have been heralded for their multiple cobenefits and as adaptive systems that buffer chronic issues associated with increasing precipitation. These systems, however, are not currently built to fully manage the more frequent extreme events that climate change may bring. New York City's fourteen wastewater treatment plants and ninety-six pumping stations, located by design at low points and along the waterfront, are particularly vulnerable to sea-level rise and storm surge. When Hurricane Sandy hit New York City in 2012, ten of DEP's fourteen wastewater treatment plants were damaged or lost power and released untreated or partially treated wastewater into local waterways. Three of these facilities were nonoperational for some time as a result of the storm—two for several hours, with one facility down for three days. The other facilities maintained at least partial treatment, including removal of pollutants and disinfection of effluent before water from these plants was discharged into waterways. Although collectively, wastewater treatment plants operated at more than twice their normal flow rate at the height of the storm, Sandy's surge led to the release of approximately 560 million gallons of untreated combined sewage, stormwater, and seawater from sewers and another approximately 800 million gallons of partially treated and disinfected wastewater into New York City waterways.

Most of the damage to wastewater facilities involved electrical systems and equipment, including substations, motors, control panels, junction boxes, and instrumentation. Sandy's floodwaters inundated the lower levels of facilities, where much of this equipment is located. Even where electrical systems were not damaged during Sandy, utility power outages forced many facilities to operate on emergency generators for up to two weeks. In addition to disrupting operations at treatment facilities, Sandy also affected pumping stations. Forty-two of ninety-six such stations were damaged or lost power. Power outages were responsible for roughly half of the impacts, with storm-surge inundation responsible for the other half, primarily in coastal communities on Staten Island and in Brooklyn and Queens. At inundated pumping stations, many of which are underground, recovery entailed not just pumping floodwaters out of the stations but also repairing damage caused by the corrosive impact of seawater on electrical equipment.

Prior to Sandy, in 2011 DEP initiated a study to understand the impacts of climate change on drainage and wastewater treatment in New York City. As part of this pilot study, DEP looked at the potential risk of storm surge and sea-level rise to one of its fourteen wastewater treatment plants. This conceptual study quickly became a reality when Hurricane Sandy struck. After Hurricane Sandy, DEP expanded its study to consider all fourteen treatment plants and ninety-six pumping stations. The study considered how to protect not only facilities damaged by the storm but also assets that could become vulnerable if a future storm were to approach from a different direction, or if sea-level rise were to increase flood heights. The result of the study is a portfolio of adaptive approaches described in the 2013 NYC Wastewater Resiliency Plan, including elevating and floodproofing critical equipment.[5] The plan showed that by investing $315 million in flood protection, DEP could avoid over $1 billion in damages if all of these assets were inundated by floodwaters. Factoring in the probability of flooding from multiple storm events, DEP estimated that the damages avoided over fifty years could amount to approximately $2.5 billion. DEP has already secured over $160 million from congressional funding that was made available after Sandy. However, since sustainable funding mechanisms are not available for resiliency, the remaining funding gap will likely have to come from DEP's own capital budget.

Amid these competing funding demands, DEP must also continue to improve infrastructure in areas of the city with limited drainage systems that currently experience street flooding and sewer backups which may be exacerbated with increases in heavy rainfall. Six of the ten wettest days on record in New York City have occurred since 1972, and this trend is projected to continue. In the southeastern portion of the Borough of Queens, DEP is developing an action plan to resolve long-standing flooding conditions that affect more than 400,000 residents. Here, flooding is caused by the interplay of several conditions, including lack of storm sewers, improperly graded streets, high impervious cover, and high groundwater levels. Further, the area is subject to tidal flooding that, with sea-level rise, may exacerbate chronic stormwater flooding conditions. The plan will consist of an intensive and accelerated long-term sewer build-out, complemented with innovative, site-specific solutions, such as bluebelts and green infrastructure, to serve as a model for other flood-prone neighborhoods of the city.

New York City is in the process of reinventing its public spaces to include stormwater management, among other uses, through interagency and public-private partnerships to improve public property using green infrastructure retrofits. One such partnership with the Trust for Public Land, a national not-for-profit open space conservation organization, is transforming old asphalt playgrounds by updating the play equipment and incorporating green infrastructure to enhance the space with rain gardens, porous paving material, and vegetation. Similarly, DEP is investing in stormwater enhancements as part of the Community Parks Initiative to improve underresourced public parks located in New York City's densely populated and growing neighborhoods with higher than average concentrations of poverty.[6] This shifting land use approach highlights the benefits of pooling resources to manage water while improving public spaces, reducing the urban heat island effect, and enhancing air quality. While the city's green infrastructure program was initially focused on reducing stormwater runoff to control combined sewer overflows, this approach of incorporating stormwater management into the design of public spaces and streets is being expanded to such areas as southeastern Queens for the purpose of reducing street flooding and controlling pollutants.

As climate change will likely continue to bring large rainfall events that exceed the capacity of drainage systems, it will become more critical to optimize land use and funding to achieve multiple objectives. Similarly, with the 2013 adoption of "A Stronger, More Resilient New York"—the city's first comprehensive coastal protection plan to reduce the risk of coastal flooding and sea-level rise—and 2015's *One New York: The Plan for a Strong and Just City*, stormwater management has entered a new realm.[7] Communities that are subject to regular flooding from heavy rainfall and high tides will likely flood more often, and areas that will be protected from flooding by increasing coastal edge elevations must ensure that there is sufficient capacity to retain rainwater behind the edge.

Conclusion

Climate change will affect water resources in New York City from the upstate watershed to New York Harbor. It will demand an innovative response by the city's water managers, planners, and regulators to meet water quality standard requirements under the Clean Water Act and SDWA while advancing the city's sustainability and resiliency objectives. As the largest municipal water utility in the United States, in a city with 520 miles of coastline and approximately 2,000 square miles of watershed, DEP must find new ways to maximize its investments by incorporating the latest climate science, affordability, and population and water demand projections, tightening regulations, and dealing with associated uncertainty. Climate change presents challenges in the form of competing funding needs and a moving target for meeting regulatory requirements. Meeting water quality criteria may become a greater challenge as rainfall increases and as the physical and chemical characteristics of water bodies shift. Furthermore, as the risks from heavy rainfall, sea-level rise, and storm surges increase, DEP will need to advance a new paradigm, one that is already in progress, for managing stormwater runoff to meet water quality, drainage, and coastal protection objectives.

DEP is integrating new tactics to manage both the extreme and chronic events, through such approaches as system optimization, emplacing green infrastructure, demand management, and providing flood protection for critical facilities. These techniques and approaches can be adapted and expanded as monitoring reveals their efficacy and as new information about climate change emerges. However, investments in resiliency and water quality continue to compete with the need to maintain a state of good repair, build new infrastructure, and invest in energy improvements to meet the city's ambitious greenhouse gas reduction goals. To balance these demands, DEP seeks flexibility from regulators regarding the distribution of resources to state-of-good-repair needs, existing and emerging regulatory mandates, and sustainability initiatives to mitigate the effects of and adapt to climate change. Moving forward, DEP will continue to pursue long-term solutions that optimize capital, maintenance, and operating costs while minimizing environmental impacts. These solutions will entail leveraging public and private partnerships, reenvisioning urban land use, improving existing infrastructure, and developing tools to optimize and enhance the existing system so that it is adaptable to a changing climate.

Notes

The authors would like to acknowledge Larry Beckhardt and David Lipsky for their contribution to the content and review of this chapter.

1. "Building the Knowledge Base for Climate Resiliency: New York City Panel on Climate Change 2015 Report," *Annals of the New York Academy of Sciences*, vol. 1336, January 2015.

2. New York City Department of Environmental Protection, *Climate Change Integrated Modeling Project* (October 2013) (www.nyc.gov/html/dep/pdf/climate /climate-change-integrated-modeling.pdf).

3. City of New York, Sustainable Stormwater Management Plan 2008 (December 2008) (http://s-media.nyc.gov/agencies/planyc2030/pdf/nyc_sustainable _stormwater_management_plan_final.pdf).

4. New York City Department of Environmental Protection, *NYC Green Infrastructure Plan* (September 2010) (www.nyc.gov/html/dep/pdf/green_infrastructure /NYCGreenInfrastructurePlan_LowRes.pdf).

5. New York City Department of Environmental Protection, *NYC Wastewater Resiliency Plan* (October 2013) (www.nyc.gov/html/dep/html/about_dep/wastewater _resiliency_plan.shtml).

6. City of New York Parks & Recreation, *NYC Parks: Framework for an Equitable Future* (October 2014) (www.nycgovparks.org/downloads/nyc-parks-framework .pdf).

7. City of New York, *A Stronger, More Resilient New York* (June 2013) (www .nyc.gov/html/sirr/html/report/report.shtml); City of New York, *One New York: The Plan for a Strong and Just City* (March 2015) (www.nyc.gov/onenyc).

Contributors

BRUCE BABBITT
Bruce Babbitt served under President Bill Clinton as secretary of the interior, in which capacity he created the first framework for interstate cooperation on the lower Colorado River. As governor of Arizona (1978–87) he negotiated the Arizona Groundwater Management Act of 1980, which remains the most comprehensive groundwater regulatory system in the nation.

ANN BLEED
Ann Bleed worked at the Nebraska Department of Natural Resources as state hydrologist and director. She helped negotiate settlements to two interstate water lawsuits before the U.S. Supreme Court and helped develop the Platte River Recovery Program and Nebraska's integrated surface water and groundwater law.

JAMES J. BUTLER JR.

James J. Butler Jr. is a senior scientist with the Kansas Geological Survey (University of Kansas). He was the 2007 Darcy Lecturer (National Ground Water Association) and the 2009 recipient of the Pioneers in Groundwater Award (Environmental and Water Resources Institute, American Society of Civil Engineers).

ALAN COHN

Alan Cohn is climate program director of New York City's Department of Environmental Protection. He leads efforts on flood protection, coordinates regional and national climate change initiatives, promotes green approaches to drainage and water quality improvement, and advances studies of climate change impacts on water supply, stormwater management, and wastewater treatment.

PAUL FLEMING

Paul Fleming directs the climate resiliency group for Seattle Public Utilities. He is former staff chair of the Water Utility Climate Alliance, the co-convening lead author of the water resources chapter of the U.S. 2014 National Climate Assessment, and active in numerous climate initiatives in the domestic and global water sector.

BURKE W. GRIGGS

Burke Griggs teaches natural resources law at the Washburn University School of Law. He has represented the State of Kansas in federal and interstate water matters, including litigation before the U.S. Supreme Court (*Kansas v. Nebraska & Colorado*, No. 126 Orig.). He was lead counsel for Kansas in negotiations over the Kickapoo Tribe reserved water rights settlement, the first of its kind in Kansas, has represented both the Kansas chief engineer and the Kansas Water Office in hearings, and advises the state's water agencies on matters of water law and policy. He is a nonresident fellow of the Woods Institute for the Environment at Stanford University and serves as an affiliated scholar at Stanford's Bill Lane Center for the American West.

KATHY JACOBS

Kathy Jacobs is a professor at the University of Arizona and director of the Center for Climate Adaptation Science and Solutions. Following three years as executive director of the Arizona Water Institute, she served in the White House as director of the National Climate Assessment and lead adviser on water science and policy and climate adaptation within the Office of Science and Technology Policy from 2009 through 2013. She has more than twenty years of experience as a water manager for the state of Arizona.

ANGELA LICATA

Angela Licata is deputy commissioner for sustainability of New York City's Department of Environmental Protection and oversees the Bureau of Environmental Planning and Analysis, the Bureau of Environmental Compliance, and the Office of Green Infrastructure. She leads initiatives on stormwater management, accelerating meaningful regulatory reform, restoring wetlands habitat, and improving air quality and public health in New York City.

JIM LOCHHEAD

Jim Lochhead is Denver Water's CEO and manager. Denver Water serves 1.3 million people in the city of Denver and surrounding suburbs. In addition to maintaining a private law practice, he has served as executive director of the Colorado Department of Natural Resources and as Colorado's representative on interstate Colorado River operations.

PAT MULROY

Pat Mulroy is a nonresident senior fellow in the Climate Adaptation and Environmental Policy unit at the Brookings Institution and a practitioner in residence at the Saltman Center for Conflict Resolution at the University of Nevada, Las Vegas, William S. Boyd School of Law. She also holds a faculty position at Desert Research. Her practice is concerned with helping communities in water-stressed areas throughout the world develop strategies to address increased water resource volatility and identify solutions that balance the needs of all stakeholders.

MAUREEN A. STAPLETON

Maureen A. Stapleton became the San Diego County Water Authority's general manager in 1996 and fashioned the agency into one of the nation's most progressive wholesale water providers. The Water Authority supports the region's 3.3 million residents and a $222 billion economy.

DOUGLAS YODER

Douglas Yoder is a local government practitioner in southern Florida, primarily in the fields of environmental regulation and utility management. Currently he is deputy director of the Miami-Dade Water and Sewer Department, the largest utility in the southeastern United States. He served for eight years on the National Drinking Water Advisory Council, currently serves on the Florida Water Resources Advisory Commission, and has been active with climate change issues since 1990.

Index